NUCLEAR FUELS POLICY

Report of the
Atlantic Council's
Nuclear Fuels Policy
Working Group

An Atlantic Council Policy Paperback

Routledge
Taylor & Francis Group
LONDON AND NEW YORK

First published 1976 by Westview Press

Published 2018 by Routledge
52 Vanderbilt Avenue, New York, NY 10017
2 Park Square, Milton Park, Abingdon, Oxon OX14 4RN

Routledge is an imprint of the Taylor & Francis Group, an informa business

Library of Congress Cataloging in Publication Data

Atlantic Council of the United States. Nuclear Fuels Policy Working Group
Nuclear fuels policy.

(An Atlantic Council policy paperback)
Includes bibliographical references.
1. Nuclear fuels. 2. Atomic power. 3. Energy policy. I. Title. II. Series: Atlantic Council of the United States. An Atlantic Council policy paperback.
TK9360.A8 1976 333.7 76-47679

ISBN 13: 978-0-367-01803-0 (hbk)
ISBN 13: 978-0-367-16790-5 (pbk)

TABLE OF CONTENTS

FOREWORD

The Atlantic Council of the United States has since 1961 been formulating policy recommendations for action on problems and opportunities shared by North America, Western Europe, Japan, Australia and New Zealand. Given the critical energy supply situation of those democratic industrialized economies, the Council has, in recent Policy Papers*, addressed the urgent issues of developing and financing alternative energy sources. This Policy Paper is focused on the nuclear energy option, and particularly on nuclear fuels supply policy.

The economic supply of nuclear energy, and therefore of nuclear fuel, to meet US and other non-Communist countries' energy demand, involves a highly complex matrix of issues. They include changing estimates of overall energy demand, other national energy alternatives, public acceptability of nuclear energy, financing, roles of the public and private sectors, the adequacy of the resource base, and the effects on national policy of international decision-making and interdependence. These are a few of the issues addressed in the course of evolving the nuclear fuel supply policy recommendations presented.

The work of the Atlantic Council is done by private citizens acting in their individual capacity, voluntarily contributing time and talent to exchange diverse, often divergent information and views, and evolving policy recommendations deemed by them to be in the national interest. Work of this nature is the backbone of the Atlantic Council's Program. This Policy Paper is a demonstration of such work. We appreciate the manner in which John E. Gray, as Chairman of the Nuclear Fuels Policy Working Group, Joseph W. Harned, as Rapporteur, and Bernard H. Cherry, as Technical Rapporteur, carried out their task. We extend our congratulations as well as our thanks to them and to each member of the Working Group.

While the final responsibility for the views expressed is that of the Working Group, the Council is pleased to present these views for public discussion and debate, which we regard as essential, as the US and the international community continue to search for solutions to these important problems.

HENRY H. FOWLER
Chairman of the Board
The Atlantic Council of the United States

**World Energy and US Leadership*, by Harlan Cleveland
Financing Free World Energy Supply and Use, by John E. Gray
Alternative Energy Sources for the US, by the Battelle Memorial Institute and Sidney E. Rolfe.

For further details on the Atlantic Council and Policy Papers Series, see page 137.

PREFACE

In the spring of 1975, the Atlantic Council's Energy Committee, under the Chairmanship of Henry H. Fowler, established a Working Group on Nuclear Fuels Policy and asked the undersigned to serve as Chairman. Subsequently, Joseph W. Harned was asked to serve as Rapporteur. We then invited appropriate members of the Atlantic Council's Energy Committee and other national and international experts to form a Working Group, and later in the course of the work asked Bernard H. Cherry to serve as Technical Rapporteur. Theodore C. Achilles, W. Randolph Burgess and Francis O. Wilcox joined us as active *ex officio* members of the Group and a small group of student interns was introduced into our work.

The members and interns of the Nuclear Fuels Policy Working Group are listed below. In addition to those listed, Philip J. Farley, Curt Heidenreich and Sidney E. Rolfe served on the Working Group during its tenure. Dr. Rolfe, a Director of the Atlantic Council and an active participant in the work of the Council's Energy Committee, served on the Working Group until his untimely death, March 10, 1976. We are also indebted to the many friends of the Atlantic Council, both in and out of government, in the US and abroad, who have provided helpful comments and suggestions.

As a Working Group, we held a series of meetings during the fall of 1975 and early in 1976 and established the outline and framework for the study. Thereafter, selected Working Group members set down in writing their views in elaboration of the various parts of the basic framework: Bernard H. Cherry, Curt Heidenreich, Ruykichi Imai, Henry D. Jacoby, W. Robert Keagy, Robert Kleiman, Harald B. Malmgren, Laurence I. Moss, L. Manning Muntzing, John Palfrey, Norman C. Rasmussen, Julian J. Steyn, and Mason Willrich. Subsequently, we prepared a series of drafts which drew largely upon the written contributions and comments and views of the Working Group as expressed in meetings or separately to the Rapporteur. In the latter stages of our work, the Rapporteur and Chairman of the Group engaged in additional consultations in the US, Western Europe, Japan and Australia, and found these to be of value in firming up the final draft for review by the Working Group. Subsequent to that review, the draft was submitted to the members of the Energy Committee of the Atlantic Council for their comment. The result of this process is presented as this Policy Paper on Nuclear Fuels.

In completing this project we would like to express, for the Council and the Energy Committee, a deep appreciation for the manner in which Messrs. Harned and Cherry and the individual members of the Working Group have made major contributions in a far-reaching and complex analysis and search for solutions. The Working Group have been unstinting in their interest and contributions, perceptive and patient in their comments and explanations. Eliane Lomax, of the Council's staff, who served throughout as our Project Assistant, provided us with good sense, good humor and hard work.

The Organization for Economic Cooperation and Development (Paris) and the Edison Electric Institute (New York) provided access to demand

and supply data available from their nuclear fuels supply studies being carried out during the same period. Financial support for research and related activities of the Working Group was made available to the Atlantic Council by the US Federal Energy Administration. (The FEA financial support does not necessarily imply endorsement by FEA of the Working Group's conclusions and recommendations.) Support for the overall Atlantic Council Energy Program, of which this effort is a part, has been made available by several US foundations, corporations, labor unions and individuals. We are most grateful for this institutional support, as well as for the extraordinary contributions of time and talent by the Working Group members.

JOHN E. GRAY
Chairman
Nuclear Fuels Policy Working Group
The Atlantic Council of the United States

NUCLEAR FUELS POLICY WORKING GROUP

Chairman:

Mr. John E. Gray, President, International Energy Associates Ltd; Director, Atlantic Council.

Rapporteur:

Mr. Joseph W. Harned, Deputy Director General, Atlantic Council; US Representative, Atlantic Institute for International Affairs, Paris.

Technical Rapporteur:

Mr. Bernard H. Cherry, Manager of Fuel Resources, General Public Utilities (GPU) Service Corporation.

Members:

Mr. David C. Acheson, Attorney; Former General Counsel, Communications Satellite Corporation; Formerly Office of the General Counsel, US Atomic Energy Commission.

Dr. Frank Barnaby, Director, Stockholm International Peace Research Institute, Stockholm, Sweden.

Mr. W. Donham Crawford, President, Edison Electric Institute.

Dr. Raymond L. Dickeman, President, Exxon Nuclear Company.

Mr. George Driscoll, Economist, National Council for US/China Trade.

Hon. William C. Foster, Former Director, Arms Control and Disarmament Agency; Former Deputy Secretary of Defense; Director, Atlantic Council.

Dr. Curt Gasteyger, Director, Arms Control and International Security Program, Graduate Institute for International Studies, Geneva; Former Deputy Director, Atlantic Institute for International Affairs, Paris and Director of Programs, International Institute for Strategic Studies, London.

Hon. T. Keith Glennan, Former Commissioner, Atomic Energy Commission; Former Administrator, National Aeronautics and Space Administration; Former Ambassador to the International Atomic Energy Agency.

General Andrew J. Goodpaster, Former Supreme Allied Commander, Europe; Director, Atlantic Council.

Mr. Michael Goppel, First Secretary, Delegation of the Commission of the European Communities to the US.

Dr. Ryukichi Imai, Deputy Manager, Engineering, the Japan Atomic Power Company; Assistant to the Chairman, Japan Enrichment and Reprocessing Group; Special Assistant to the Foreign Minister; Consultant to MITI.

Hon. John N. Irwin, Attorney; Former Assistant Secretary of Defense, Deputy Secretary of State and Ambassador to France; Director, Atlantic Council.

Dr. Henry D. Jacoby, Professor of Economics, Sloan School of Management, Massachusetts Institute of Technology.

Mr. W. Robert Keagy, International Energy Consultant, Zurich.

Mr. Robert Kleiman, Editorial Board, *The New York Times*.

Hon. John M. Leddy, Former Assistant Secretary of State and Treasury; Former Ambassador to the OECD; Director, Atlantic Council.

Hon. David E. Lilienthal, Chairman, Development and Resources Corporation; Former Chairman, Atomic Energy Commission.

Hon. Harald B. Malmgren, Woodrow Wilson Fellow, Smithsonian Institution; Former Deputy Special Representative for Trade Negotiations; Director, Atlantic Council.

Mr. R. Eric Miller, Vice President, Bechtel Corporation.

Mr. Laurence I. Moss, Consultant on Energy/Environment; Chairman, Federal Energy Administration's Environment Advisory Committee; Former Executive Secretary, Committee on Public Engineering Policy, National Academy of Sciences; Former President, Sierra Club.

Mr. L. Manning Muntzing, Attorney; Former Director of Regulation, Atomic Energy Commission.

Hon. Paul H. Nitze, Former Delegate to SALT; Former Deputy Secretary of Defense and Secretary of the Navy; Director, Atlantic Council.

Dr. John Palfrey, Professor of Law, Columbia University; Former Commissioner, Atomic Energy Commission.

Dr. Jean-Pierre Poullier, International Economist, Paris.

Dr. Norman C. Rasmussen, Professor of Nuclear Engineering and Head of the Department of Nuclear Engineering, Massachusetts Institute of Technology.

Hon. John C. Sawhill, President, New York University; Former Administrator, Federal Energy Office.

Dr. Julian J. Steyn, Vice President, NUS Corporation.

Dr. Mason Willrich, Professor of Law, University of Virginia; Former Assistant General Counsel, Arms Control and Disarmament Agency.

Project Assistant:

Mrs. Eliane Lomax, Atlantic Council.

Ex-Officio Members:

Hon. Theodore C. Achilles, Vice Chairman, Atlantic Council; Former Counselor of the State Department and Ambassador to Peru.

Hon. W. Randolph Burgess, Vice Chairman, Atlantic Council; Former Under Secretary of the Treasury and Ambassador to NATO and the OEEC.

Dr. Francis O. Wilcox, Director General, Atlantic Council; Former: Executive Director, Commission on the Organization of the Government for the Conduct of Foreign Policy; Assistant Secretary of State; Chief of Staff of the Senate Foreign Relations Committee; Dean of Johns Hopkins School of Advanced International Studies.

Student Interns:

Mr. Kevin Conry, Georgetown University, Washington, DC.
Mr. Ross Hoff, Georgetown University, Washington DC.
Miss Keiko Itoh, Sophia University, Tokyo, and Yale University, New Haven.

INTRODUCTION AND SUMMARY

The purpose of this Policy Paper is to recommend the actions deemed necessary today to assure that future US and other non-Communist countries' nuclear fuels supply will be adequate to meet future energy demand. Taken together, the recommended decisions and actions form *a nuclear fuels supply policy* for the United States Government and for the private sector, and new areas of responsibility for the appropriate international organizations in which the US participates.

The recommendations have been developed in the light of the following considerations:

- estimates of modest growth in overall energy demand, electrical energy demand, and nuclear electrical energy demand in the US and abroad, predicated upon the continuing trends involving conservation of energy, increased use of electricity and moderate economic growth (Chapter I);
- possibilities for the development and use of *all* domestic resources providing energy alternatives to imported oil and gas, consonant with current environmental, health and safety concerns (Chapter II);
- assessment of the traditional energy sources which provide current alternatives to nuclear energy: domestic oil and gas, hydro and geothermal energy, and, in particular, coal—the most abundant fossil fuel resource currently available in the US (Chapter II);
- evaluation of realistic expectations for additional future energy supplies from prospective technologies: enhanced recovery from traditional sources, the development and use of oil shales and synthetic fuels from coal, fusion and solar energy (Chapter II);
- an accounting of established nuclear technology in use today, in particular the light water nuclear power reactor, used for generating electricity (Chapter III);
- an estimate of future nuclear technology, in particular the prospective fast breeder reactor generation (Chapter IV);
- current and projected nuclear fuel demand and supply in the US and abroad (Chapters V and VI);
- the constraints encountered today in meeting nuclear fuels demand (Chapter VII);
- and the major unresolved issues and options in nuclear fuels supply and use (Chapter VIII).

The results of these assessments are specific recommendations for action by the US Government and the private sector, as set out in Chapter IX.

The principal conclusions and recommendations are that the US and the other industrialized non-Communist countries should strive for increased flexibility of primary energy fuel sources, and that a balanced energy strategy therefore depends upon the security of supply of energy resources and the ability to substitute one form of fuel for another. The substitutability and efficient use of energy resources are enhanced by accelerating the supply and use of electricity.

In the US, electric utilities must shift from burning increasingly scarce and costly oil and gas. The production of coal can and should be expanded within the limitations of mining and burning it in accordance with evolving health, safety and environmental regulations, and the logistical constraints of delivering it. Other alternatives—wind, water power, geothermal heat, sunshine and fusion—are not seen as likely to generate significant amounts of electricity for decades.

A strong effort to promote more rigorous energy conservation is needed and should result in more efficient energy use. But even with more efficient energy use, North American, European and Japanese needs for energy are bound to increase with the growth of population and increased economic activity. Thus the US, like other leading industrialized nations, really has no choice but to use all of its domestic energy options. And, at least for the next three or four decades, the mix must include an increasing amount of nuclear fuel for light water reactors.

The US should therefore reassert its traditional leadership in international nuclear cooperation by promptly assuring adequate domestic and interdependent non-Communist countries' nuclear fuels supplies under effective guarantees and safeguards. And the design and implementation of an effective non-proliferation policy should therefore continue to be a high national security goal for the US, today and in the foreseeable future.

US initiatives in international nuclear cooperation should be directed to achieving the following goals:

- assurance of adequate US and other non-Communist countries' energy supplies;
- support of the efforts of the nuclear supplier nations and the parties to the Non-Proliferation Treaty (NPT) to stem further proliferation of nuclear explosive devices; and
- relief from the pressures on the US and other non-Communist countries for increasing dependence on imported oil.

In order to accomplish these goals, the US must rapidly expand domestic nuclear fuel supply capacity and attendant technology, as follows:

a) increase domestic uranium production capacity;

b) expand domestic uranium reserves;

c) expand US enrichment capacity;

d) demonstrate US reprocessing and recycling technology, economics, and safeguards, as a prerequisite to commercial operation;

e) accelerate the development of effective radioactive waste management; and

f) accelerate the further development and implementation of safeguards to deter, prevent and detect diversion of nuclear materials, facilities and technology to other than peaceful purposes.

The expansion of domestic nuclear fuel production capacity and technology recommended above should be financed using to the maximum extent the sources of capital found in domestic and foreign capital markets. Where private capital is not forthcoming promptly, governmental financial guarantees should be considered. This principle should apply to the financing of all components of the nuclear fuel cycle requiring rapid expansion in the US. The principle might usefully be extended to cover capitalization of multinational and bilateral nuclear fuel supply ventures.

The Government should accelerate its current schedule for re-evaluating natural uranium reserves, in order that sound supply planning may proceed. It should as well promptly remove all restrictions to the import of uranium to the US. Purchases of uranium from all currently producing and potentially exporting countries should be considered. It may also be advisable to consider a uranium inventory policy aimed at reducing possible concerns about over-reliance on foreign uranium sources.

The utility industry, in collaboration with the potential commercial reprocessors, should assure adequate recoverable storage of spent fuel elements until commercial reprocessing becomes available. The Government should in parallel develop a contingency plan which could result in spent fuel being accepted as the Government's responsibility for further storage and ultimate disposal.

The Government should proceed to cooperate with and encourage industry to demonstrate nuclear fuel reprocessing and recycling technology, economics and safeguards on a commercial scale, in order that decisions as to the further commercial development of reprocessing may be made on a sound and logical basis. The Government should continue to safely store and implement plans for the permanent and secure disposal of existing high level radioactive wastes, and should have available an adequate program for the permanent secure disposal of prospective high level radioactive wastes.

The US Government and private sector should jointly explore with the traditional and potential foreign buyers of US enrichment services the possibility of exporting US enrichment technology, and eventually reprocessing technology, to multinational and bilateral nuclear fuel supply centers, in order to take full advantage of the prospective economies of scale and supply security such centers would provide, while at the same time safeguarding the further proliferation of enrichment and reprocessing technology. If and when reprocessing technology is added to multinational and bilateral nuclear fuel supply centers, consideration should be given to co-locating reprocessing and fuel fabrication facilities at those centers, thereby enhancing international safeguards.

The International Atomic Energy Agency (IAEA, Vienna) should be provided adequate funding and manpower, as well as access to national data and facilities, as more and increasingly sensitive facilities become subject to IAEA standards. IAEA accountability should minimize time lags in learning of possible diversion of special nuclear materials. National verification, accountability, and physical security knowhow should be made available to and pooled through the IAEA.

The IAEA should be provided adequate funding and manpower as well to ensure an appropriate active role in developing safety, environmental and health standards for multinational and bilateral nuclear fuel supply centers. The Agency, in cooperation with the other appropriate international organizations, should act as an international clearinghouse:

a) for national data on the projected demand for nuclear fuel supply services in all of the developing and the industrialized countries of the world, and

b) for the exchange of information and the exploration of possibilities for sharing technology concerning the various national programs to develop fission breeders and fusion reactors as long-term future energy sources.

Sensitive international nuclear fuel cycle facilities should be co-located to eliminate or reduce vulnerable transportation links. Safeguards measures to be applied during international transportation of special nuclear materials should be standardized insofar as possible. International standard measuring systems should be established in order to facilitate the accountability and verification of nuclear materials. Maximum inspection frequencies and access rights should be required of importers of nuclear materials and knowhow.

The US and other NPT proponents should make a determined effort to bring additional nations into the NPT. A new international agreement or treaty should be established to provide for the physical security of nuclear materials and facilities.

Until all supplier nations are party to the NPT, the recently expanded *ad hoc* "London Group" of supplier nations should continue to act as the informal bridge to substantive agreement on rigorous safeguards triggered by supplier country exports. A common policy should be agreed upon by all supplier nations to establish strict standards governing the export of sensitive nuclear materials and technology and to assure their enforcement on a case-by-case basis.

International depositories for plutonium in excess of peaceful national needs—which are provided for in the IAEA Statutes but have not yet been established—should be created now to accept and safely store plutonium under the aegis of IAEA. Such international depositories should include provision for the storage of spent fuel containing plutonium.

Nuclear Fuels Policy

I. NUCLEAR ENERGY — WHO NEEDS IT?

Electrical energy has been and will continue to be a significant part of growing world energy requirements. In the face of diminishing reserves and increased costs of fossil fuels, the use of nuclear fuel to produce electricity has emerged as a practical and economic alternative. The following analysis of the outlook for nuclear power in the US and abroad provides a preliminary view of why nuclear power is important to national energy planning. Foreign nuclear fuel demand, supply and use are considered in this Policy Paper to the extent that they are likely to continue to affect US policy decisions.

A. The Outlook for the United States:

All methods of meeting the energy demands of the US are severely constrained, according to the US Energy Research and Development Administration (US ERDA):[1]

- "Domestic crude oil production peaked in 1970 and has declined by more than one million barrels per day since then. Production is now at a nine-year low.
- "Oil imports are about 37 percent of oil consumption and would rise to more than 50 percent of consumption of 12 million barrels per day by 1985 if no new actions are taken.
- "As a result of increasing import dependence, US payments to foreign producers for imported oil increased from less than $3 billion in 1970 to about $27 billion in 1975 and will increase by another $2 billion annually, largely because of the OPEC price increases.
- "Natural gas production peaked in 1973, declined by 6 percent in 1974 (the equivalent of over 230 million barrels of oil), and dropped another 8.5 percent during the first half of 1975.
- "Electric utility financial problems and regulatory delays have in part

[1]"Energy Fact Sheet", The White House, October 10, 1975.

resulted in the cancellation or postponement of about three-fourths of all planned nuclear plants and about one-third of all coal plants previously scheduled to come into operation between now and 1985.

- "Some emerging technologies, such as synthetic fuels from coal, shale oil, solar, and methods to use energy more efficiently, have uncertain economics due to long lead times and technological uncertainties, and considerable risk if world oil prices should drop."

Despite the magnitude of these difficulties, however, the US (along with Canada) is better off than other industrial nations in that it has alternative energy options based on large indigenous coal and uranium supplies while most of the others do not.

1. Total Energy Demand

With the rapid worldwide rise in energy demand, quadrupling in Japan, tripling in the European Community and doubling in the US between 1950 and 1970, the US is still the big energy consumer, using twice as much per capita as the British, 2½ times as much as the Germans, and 4½ times as much as the Japanese.

Total energy demand forecasts for the US through the end of the century are given in *Table I.1*[2]. These forecasts are consistent with recent projections of both the US Government and the international Organization for Economic Cooperation and Development (OECD). The projections indicate that total energy demand in the US will continue to grow at the rate of about 3 to 3.5 percent per year through the remainder of this century.

While energy demand in the US will depend upon economic growth, per capita energy utilization, conservation and other factors difficult to predict such as population growth, the 3 to 3.5 percent range used here for the US is also consistent with the evolving consensus that energy demand for the period through 1985 will be substantially larger than current demand.

2. Electrical Energy Demand

In *Figure I.1.*, recent US ERDA projections of total US installed electrical generating capacity are presented along with those of the private National Electricity Reliability Council (NERC) made in 1974 and 1975. It can be seen that the most recent NERC forecast is more conservative than that made a year earlier and that the ERDA forecast is even lower, a 20 percent reduction by 1980 and approximately 25 percent reduction by 1985 from the forecasts made a year earlier. Electrical growth rates are now projected to be in a range from 4.75 to 6.40 percent annually.

The ERDA projections of the growth in total electrical energy generating capacity in the US are broken down by fuel option in *Table I.2*. It can be seen that the estimated range for 1985 is from 785,000 to 875,000 megawatts of electrical generating capacity, compared with 500,000 megawatts in 1975.

If oil importation is to be held constant, or hopefully reduced, and domes-

[2]All tables and figures not presented in the text are found in Appendix A.

tic gas supply is diminishing, then a combination of efforts, including conservation, efficiency improvements, and accelerated electrification based on non-petroleum fuels are necessary. In spite of the approximately zero growth in demand for electricity in 1974, a year of recession, it could be as high as 5 to 6 percent annually over the next ten to fifteen years.

3. Why Electrification?

The reasons for enhanced electrification based on coal and nuclear fuel, as presented in a recent Atlantic Council Policy Paper[3], are the following:

- Electricity is economical, versatile, and reliable. For these reasons its increased use in the US has been a persisting historic trend.
- Acceleration of this trend, using nuclear fuel and coal to generate electricity, provides the opportunity to:
  - — cut back on US oil and gas requirements and their associated political, monetary, and international supply problems;
  - — develop a US energy policy and supply based on resources and decision-making under US control;
  - — use all US controlled energy resources more efficiently.
- An expanding electricity supply program is feasible—with adequate financing and government/industry cooperation and coordination in setting goals and decision-making.
- An expanding electricity use program is feasible—with encouragement, education, and incentives in the residential, commercial, industrial, and transportation sectors.

In 1973, 26 percent of all primary energy used in the US was converted to electricity. 1973 (pre-oil crisis) estimates were that by the year 2000 from 50 to 65 percent of all primary energy used in the US would be used in the form of electricity.

Accelerating the supply and use of electricity, using nuclear and coal fired plants, would provide the opportunity to:

- Phase out the use of oil and gas for generation of electricity.
- Substitute electricity for oil and gas in many of their end use energy applications.
- Provide earliest and maximum US long-term independence of interruptible and high cost imported oil and gas.
- Develop US energy policy compatible with conservation and environmental requirements.
- Encourage use of an energy form which in many major applications, such as heating and some forms of transportation, can be more efficient in the use of energy resources (BTUs) than alternate energy sources (i.e., the efficiency losses in generating electricity are more than offset

[3]*Financing Free World Energy Supply and Use;* John E. Gray; POLICY PAPERS: Energy Series; Atlantic Council of the United States; February 1975; pages 8-10.

by the higher efficiency in the end use resulting from the use of electricity, presuming correct end use).

- Use currently established institutions, technology and capabilities which are sufficient to quickly accelerate electricity supply and use.
- Develop and incorporate into an ongoing program alternate energy sources—fusion, solar, geothermal, tides, winds, etc., as these sources are developed and can contribute economically to the generation of electricity.
- Conserve remaining US oil and gas reserves for uses in which they have unique value—some transportation (including fewer, higher efficiency automobiles; airplanes and trucks); some electric generation (such as very low load factor peaking plants); and products such as fertilizers, petrochemicals, plastics, drugs, etc.

Over the period since World War II, electric power consumption grew in most countries at roughly twice the rate of energy consumption as a whole. The increased physical scarcity of oil and gas (apart from the politically created scarcity), the costs of converting solid hydrocarbons into liquids or gases for consumption, and the availability of coal and nuclear fuel for power production may accelerate the electrification trend, even though overall energy growth rates may slow somewhat.

4. Nuclear Electrical Energy Demand

The projection of nuclear electrical energy demand has been uncertain, as evidenced by the various forecasts made in the US over the past ten to fifteen years, as borne out in *Table I.3*. Changes in projections have been due to slippage in schedules and cancellation of commitments by the electric utility industry. These changes have been due to the difficult financial conditions encountered by the utility industry in the recent past, as well as to overall energy demand uncertainties, nuclear fuel supply difficulties, regulatory obstacles, and public acceptance problems which have adversely affected utility planning. The forecasting experience in Western Europe, Japan and other countries has been similar, and is treated later in this chapter.

Nuclear power growth forecasts for the US are set forth in *Tables I.4* and *I.5*. An examination of the forecasts indicates that nuclear power growth rates of the order of 12 to 16 percent per year are predicted for the US over the next twenty years, in the case of both the high and the moderate overall energy growth scenarios developed by the US ERDA[4]. In both cases, US nuclear power capacity is predicted to approximately double in intervals of less than five years until the early nineties. In the lowest overall energy growth scenario, nuclear power growth is forecast to be 14 percent per year

[4]Nuclear power growth forecasts given in this Policy Paper are based upon data developed by the US ERDA, the International Atomic Energy Agency (IAEA), the Organization for Economic Cooperation and Development (OECD), and the US utility industry provided through the Edison Electric Institute (EEI) Nuclear Fuels Supply Study Program.

in the period 1980-85, but decreasing dramatically thereafter.[5] In order to properly understand the forecasts for the next ten years, it should be realized that they are based on plants which are either already operating, under construction, or largely committed today.

The deferrals and cancellations of nuclear plants in the US over the past two years serve to lend some uncertainty in nuclear power growth beyond about 1983. Deferrals in other countries have resulted in a similar though somewhat less accentuated situation for the period after about 1981. While it is possible to conceive of accelerated nuclear programs in the near future, the opposite is also imaginable, especially in the US where actions such as those of the California Initiative could have widespread effects.

While current decision-making to meet the future demand for nuclear fuel supply will not be significantly influenced by 10 to 15 percent uncertainties in the 1985 demand, it would of course be completely altered if no further growth in nuclear power were to be the case in the remainder of this decade. Put another way, the 1985 demand projected for nuclear fuel may be considered as only being certain to within about one year either way, assuming an orderly course of events.

5. Why the Nuclear Demand?

Even if the US accelerates electrification, develops domestic energy supply to the maximum, and increases energy end use efficiency, ERDA and other studies indicate that the US will be a net importer of energy through 1995[6]. Thus, any decrease in the expansion of coal and nuclear electrical power generation development will increase the degree to which the US is dependent upon imports to fill its energy needs.

Development of coal resources is already considered as the concomitant and not an alternative to nuclear energy generation. The selection of nuclear electric generation as the complement to coal results from domestic availability of uranium and its cost and reliability.[7] Current domestic uranium resources are estimated as more than capable of supporting existing and committed nuclear plants. Expansion of known uranium reserves is considered promising geologically. Problems associated with this expansion are discussed in later sections of this paper. Other less developed sources of energy, such as solar, winds and fusion, cannot be counted upon to make a significant contribution to total energy needs in the next 15 to 20 years.

Nuclear fuel is now economically attractive compared to fossil fuels as a means of generating electric power. Nuclear fuel cost advantages over

---

[5]Analysis resulted in the determination of a high, moderate and low forecast through the firm commitment period of the mid-eighties and thereafter. The high, moderate and low forecasts are given in *Table I.4;* they correspond to the forecast curves shown graphically in the lower portion of *Figure I.1.* Forecast of foreign non-Communist and Communist nuclear power growth projected in 1975 by US ERDA were obtained through the EEI; they are presented in the table along with world total data.

[6]*A National Plan for Energy Research, Development and Demonstration: Creating Energy Choices for the Future;* US ERDA; two volumes (ERDA-48); Washington, DC; June 1975.

[7]For an understanding of how decisions are made on which type of electrical generation, see Appendix B.III.

coal-produced energy are as high as 5:1, according to recent experience in the Northeastern United States. US nuclear power plants generated electricity in the first half of 1975 at 43.6 percent less total cost than fossil fuel (oil/coal) plants. This represented savings of $670 million for that half-year. It also means fossil fuel savings equaling 115 million barrels of oil or 25 million tons of coal. Despite prolonged outages of several US nuclear power plants during the first half of 1975, their overall reliability for the period compared favorably with that of conventional generators. The average capacity factor for domestic nuclear plants in the period was 64.2 percent compared to 51.9 percent for domestic base-loaded plants using fossil fuels.

B. The Outlook Abroad

1. Demand and Problems

There remains a marked difference in the energy supply vulnerabilities among the industrial nations: OPEC accounts for roughly three-fourths of Japan's primary energy supply, and two-thirds of Europe's, as compared to 17 percent of US primary energy supply. The difference is the result of the relatively favorable domestic energy resource base enjoyed by the US; in comparison with the US, Japan and most of Europe have less significant domestic energy alternatives to imported oil and gas.

However, it is the non-oil producing developing countries of Africa, Latin America and Asia, and their economic development plans, which are the hardest hit of all in the world energy crisis. Their balance-of-payments problems, always difficult, have now become acute.

The Communist countries have also had to reassess their potential for substituting domestic energy resources for imports, accelerate their energy research, development and demonstration efforts on alternative energy sources, and begin to expand rapidly their already substantial programs in hydro and nuclear electric generation. The Soviet Union and the People's Republic of China both have expanded energy exports. The USSR has stepped up contracts to supply natural gas and uranium enrichment services in Western Europe, is looking to sell oil to the US, and is charging higher prices for energy to its East European partners. China is beginning to export significant amounts of oil and coal to Japan and other Asian countries.

Forecasts of total energy demand for the OECD nations (Western Europe, North America, Japan, Australia and New Zealand) are given in *Table I.6* on the basis of oil prices in 1980 and 1985 (in 1974 dollars). These data suggest that the growth rate of total world energy demand will approach that projected above for the US, in the range of 3 percent.

Nuclear power growth forecasts for the US and Canada, the European Community, the European OECD countries, Japan, Pacific Asia and the Communist countries are set forth in *Table I.5*. An examination of the forecasts indicates that nuclear power growth rates on the order of 14 to 16 percent per year are predicted both for the US and the world.

2. The Reality of Nuclear Power

An idea of how extensive the spread of nuclear power technology around the world has become, and of how rapidly it is likely to continue, can be found in the following world picture.

At the end of 1975[8], 168 nuclear power reactors with a generating capacity of about 73,000 MWe were operating in 19 countries (Argentina, Belgium, Bulgaria, Canada, Czechoslovakia, France, Federal Republic of Germany, Democratic Republic of Germany, India, Italy, Japan, the Netherlands, Pakistan, Spain, Sweden, Switzerland, the UK, the US and the USSR).

All of these countries except the Netherlands and Pakistan have additional commercial power reactors under construction. China is said to have constructed one or two power reactors. An additional nine countries have their first commercial power reactors under construction: Austria, Brazil, Finland, Hungary, Iran, Mexico, South Korea, Taiwan and Yugoslavia. And many others including Bangladesh, Cuba, Egypt, Indonesia, Iraq, Israel, Kuwait, Libya, Luxembourg, the Philippines, Poland, Romania, Saudi Arabia, South Africa, Thailand and Turkey have announced plans to acquire power reactors.

By 1980, if present plans are carried through, 29 countries will have installed nuclear power reactors with a total electrical generating capacity of about 219,300 MWe, about eleven times the 1970 figure. The potential world market for nuclear plants from 1980 to 1990 is summarized in *Table I.7* as estimated by the World Bank (IBRD) and the International Atomic Energy Agency (IAEA).

Given the current and projected worldwide distribution of nuclear power reactors, practically every nation is today interested to some significant degree in obtaining or developing supplies of nuclear fuels. Reasons for this interest include energy demand, security and independence of supply, economic benefit, and the cost of alternative energy sources.

The reality of nuclear power in world energy supply planning is demonstrated in *Table I.8* and *I.9*. These two tables, compiled by SIPRI in 1976 and 1975, respectively, show what nations have embarked on nuclear power programs and what nations have developed, or are developing, nuclear power plants and nuclear fuels supply capabilities.

Table I.8 indicates the acquisition or development, country by country, of nuclear power reactors; enrichment and reprocessing capabilities; uranium supplies; research reactors; fast breeder reactor development; NPT status and other Agreements; and membership in the various international organizations as of January 1, 1976.

Table I.9 lists, country by country, the number of nuclear power reactors; dominant reactor type; total nuclear electric power output produced; approximate annual production and stocks of plutonium; uranium resource estimates and planned production; reprocessing facilities; enrichment plants and plans; breeder reactor developments; and nuclear weapons status as of January 1, 1975. What these two tables indicate is that there has been a

[8]This description of the spread of nuclear technology is presented as reported by the Stockholm International Peace Research Institute (SIPRI) in the 1976 revision and update of *The Nuclear Age*, using data provided by the International Atomic Energy Agency (IAEA).

dramatic increase in installed nuclear generating capacity, significant firming-up of nuclear fuels supply planning, and concomitant attention paid by the world community to the attendant proliferation problems.

II. ALTERNATIVES TO NUCLEAR ENERGY

A. Traditional Energy Sources

1. Oil and Gas

The major feature of the energy policies of Western industrialized nations which led to the energy crisis was and is an overdependence on oil and natural gas. None of these nations, including the US, is self-sufficient in these fuels. Oil and natural gas currently supply more than three-quarters of the current energy needs of the industrialized world. Utilization of domestic petroleum resources is already constrained by available supply. US imports supply approximately 40 percent of its petroleum consumption. It is estimated that US known and recoverable oil and natural gas reserves including Alaskan oil, if used at the current rate, will be depleted in about 60 years[9]. The US has already utilized almost half of its potential petroleum resources.

On a worldwide basis, the situation is not much different. Current estimates of probable and undiscovered potential petroleum reserves are about two trillion barrels. While reserves continue to be revised upward, at projected rates of production this reserve will probably be depleted in about 80 years[9]. Careful use of such a depletable resource is in the interest of the world community.

2. Coal

Coal is the most abundant fossil fuel resource currently available in the US. US coal reserves are on the order of three trillion tons, more than half of which have been mapped and explored, and more than one-quarter of which are recoverable using current technology. Reserves are clearly not a limiting factor in the utilization of coal. Current demand estimates indicate a required production level of more than one billion tons of coal annually by 1985. This represents a substantial increase from current production levels of approximately 600 million tons. Additionally, the movement of these increased quantities of coal will require major improvements in and increases in the capacity of existing railroad transportation systems, as well as the introduction of other means of transporting coal, such as coal slurry pipelines. Current estimates of industry capacity to meet this projected demand are not encouraging. While coal industry sources (National Coal Association) indicate possible expansion to required levels, other industry ob-

[9] The specific projections used above were extrapolated and developed by Bernard H. Cherry from US and world data contained in ERDA-48, *A National Plan for Energy Research, Development and Demonstration: Creating Energy Choices For The Future*, augmented by the data presented by Robert D. Moody and Robert E. Geiger in the "MIT Technological Review", March-April 1975. See also: *Alternative Energy Sources for the United States*, by Richard Anderson, Peter Hofmann and Sidney E. Rolfe; Atlantic Council, 1975.

servers are not as optimistic. Potential constraints on expansion fall into the following areas:

a) Required Capital—expansion to target levels will probably require an additional 25-30 billion dollars of investments in an industry which is currently capitalized at 5 billion dollars.

b) Required Manpower—an additional 125,000 miners and 10,000 mining engineers are required for an industry which is having trouble meeting current demands.

c) Impact of Health and Safety Regulations—decreased productivity due to more rigorous application of mine health and safety legislation is expected to continue.

d) Impact of Unions—the impact of an increasingly powerful coal miners' union may affect productivity as the industry expands.

e) Impact of Strip Mining Legislation—delays in the expansion of the industry due to uncertainties surrounding proposed strip mining legislation are likely to affect the industry's ability to expand.

f) Transportation Requirements—expansion to target levels will require the substantial upgrading of railroads, particularly those in the Northeastern US, and maximum utilization of alternative coal slurry pipelines.

g) Impact of Air Pollution Regulations—there are major unresolved issues as to the allowable discharges from coal burning electric generating stations and the facilities (scrubbers) to be used in controlling discharges.

The factors noted above, and the currently uncertain status of development of Western coal reserves due to environmental litigation, make the ability of the industry to expand to projected levels by 1985 subject to doubt. Further, the industry's ability to attract the large amounts of required capital is not assured. The coal industry has not been viewed favorably by the investment community in the past due to uncertain profits and less than aggressive management.

In summary, the expansion of coal production capacity to projected levels is open to doubt. And even though overall coal supply and demand may be brought into balance, since the bulk of increased coal production is likely to be from Western US coal reserves, the Eastern user will be in jeopardy of shortfalls in supply. The imposition of a substantial demand increase on the coal industry to provide, additionally, the increments of primary energy expected from nuclear fuel would place an enormous burden on an industry already under strain, particularly in the Eastern United States.

3. Hydropower

In 1971 hydroelectric generation amounted to 4.0% of US total primary energy consumption: by 1975 it had increased to 4.4%. Hydroelectric generation is projected to decline to 4.2% of primary energy consumption in 1980, 3.7% in 1990 and 3.1% in the year 2000[10]. While hydropower can contribute

[10]See in Appendix A: *Table II.1.* "US Total Gross Consumption of Energy Resources to the year 2000", originally presented in the Atlantic Council's *Policy Paper* "Alternative Energy Sources for the US", by Richard Anderson, Peter Hofmann and Sidney Rolfe; April 1975.

only modestly to total US primary energy needs, it should be further exploited.

The above projections of the share of total primary energy provided by hydropower assume higher utilization of US hydro potential, consonant with agricultural, environmental and other public policies. Only about 60% of that potential has been harnessed to date. However, it is clear that full utilization of the potential of hydropower would not significantly decrease the need for coal and nuclear power generation. Further, in most areas of the US, available sites for hydropower are limited.

4. Geothermal Energy

Geothermal energy is another potential energy resource. In the United States, the use of natural steam to generate electricity is limited to The Geysers area in north-central California, where a 300 megawatt station is operated by Pacific Gas and Electric Company. Several geothermal electric stations are also in operation in other countries. Geothermal sources can be broadly divided into three categories: vapor-dominated, liquid-dominated, and dry-hot-rock systems. Vapor-dominated reservoirs, e.g., The Geysers, are most suitable for commercial power exploitation, but their occurrence is quite rare. Liquid-dominated systems are more common in the United States, but the technology of utilizing low-temperature water for useful purposes is not well developed. Finally, the dry-hot-rock systems could potentially represent a very large reservoir of energy. However, the technology for exploiting these reservoirs is in its infancy. In any event, an optimistic view of the prospective contribution of geothermal services in the US to the production of electricity is 2 to 3% in 1990.

B. Prospective Corollary Energy Sources

The addition of resources to existing fossil reserves through the development of enhanced recovery techniques and the development of corollary sources is necessary. While none of these is expected to provide significant relief from energy demands on existing resources, together they would help to buy the time needed to develop new energy sources for the future.

1. Enhanced Recovery from Traditional Sources

Enhanced recovery from existing oil reserves, while dependent on the price of oil, could provide for recovery of up to an additional 40 billion barrels. Additional technological development is required for recovery techniques along with additional investment. If the technology and the capital are made available promptly, the additional oil developed through this approach could extend the time available to the US to gear up other energy supply sources. The limiting factors are those germane to all energy development: technical feasibility, economics, environmental impact and investment requirements.

2. Oil Shales

Oil shales are a very large and very special type of energy resource available in huge quantities in northwestern Colorado, northeastern Utah, and southwestern Wyoming. The carbonaceous substance contained in the rock is a material known as "kerogen", a brownish-black material that will melt at 600°F, producing a sticky fluid which reverts to a solid as soon as the temperature drops. Kerogen-rich layers of rock are exposed in canyon walls in some of the most spectacular mountain scenery in the West. The problems, which exploitation of oil shales face, are many: disruption of the surface, pollution from kerogen refining, disposal of waste rock, provision of an adequate water supply, and ability of the area to support sudden increases in population, to name a few.

Estimates of ultimately recoverable oil from shale vary depending upon assumptions of available technology and economics. Shales vary considerably in their oil content per ton, and recovery techniques are not firmly established. If technology could be developed to reach oil economically from the more dilute shales (containing between ¼ and ½ bb per ton of shale), the oil shale resource base could be twice the size of our remaining oil and gas reserves.

Given a favorable outcome of the technological, environmental and economic problems associated with shale-oil recovery, it could make a significant contribution in the late 1990's or early in the 21st century. The timing of such a contribution is dependent upon some of the same factors which affect the expansion of the coal mining industry, except that oil shale is starting from a much smaller base and with major uncertainties regarding the economics of recovery.

3. Synthetic Oil and Gas from Coal

Coal gasification and liquefaction represesent other approaches toward replacing oil and natural gas fuel. The production from coal of either liquid or gas fuel is not a new art. Most of the basic technology was developed in the 1930's and 1940's. Revisions of these methods and processes, as well as new technologies, are now needed. Many concurrent investigations of coal conversion processes are in progress, all of which seek an improved method for making an acceptable gas or liquid fuel such as methanol. Some are being supported by Federal research funds, others by industry, and a few represent joint undertakings. Two important characteristics of all of these processes are the substantial capital required, and the long lead times projected before commercial production can begin on a scale large enough to make a significant contribution to our energy supply.

C. Future Energy Technologies

While the traditional energy sources (gas, oil, coal, hydro and geothermal) and their prospective corollaries (oil shale, coal-based synthetics) involve an array of established and developing technologies, two sources of energy, solar and fusion, involve new technological attempts to take more direct advantage of the primary source of energy for all life: the sun. Solar displays are designed to collect the sun's fusion energy at a distance of 93 million miles, while other attempts are being made to bring the sun to earth by developing controlled and sustained fusion reactions managed by man.

1. Fusion[11]

Much effort is being devoted to producing a controlled fusion or thermonuclear reactor. Fusion is the ultimate source of the energy of the sun and the stars—in which nuclei of hydrogen are constantly fusing together to form helium. In the interior of these bodies, temperatures range from 5-100 million degrees Celsius, sufficient for fusion to occur.

Stellar fusion reactions take place only at a slow rate. Consequently, to produce large amounts of energy in a short time even higher temperatures than those found in the stars are required. But no existing solid material can withstand temperatures higher than a few thousand degrees. And the major problem in producing a fusion reactor, which so far has not been solved, is to isolate the fuel from the walls of its containment vessel.

Several countries are performing fusion research, including the USSR, the UK, France and FR Germany in addition to the US. The EEC countries are coordinating their research through Euratom.

There is considerable difficulty in determining when fusion reactors might make a significant contribution to energy supply. Historically the time from the development of a new energy source to its use on a wide scale has taken 60 years. In the case of nuclear power, this time is on the order of 40 years. This would indicate that, considering this factor alone, the commercial utilization of fusion reactors would be unlikely before the year 2015. The problem with planning on fusion as a major new energy source is lack of proof of scientific or technical feasibility and the inability to develop realistic plans and estimates until technical and then engineering feasibility is established.

2. Solar Energy[12]

Solar energy is plentiful. But solar energy is also a very dilute and periodic energy source. Conversion of solar energy to electrical energy or to some other useful energy form presents no insurmountable technical problem. To accomplish this in a reasonably economic fashion is another matter.

Solar energy should be considered as encompassing all types of energy derived from the present activity of the sun. From this perspective, the following are types of solar energy application: direct utilization of sunlight, such as heating and cooling of buildings and process-heat systems; electric power generation, including photovoltaic conversion, solar thermal conversion, wind (and hydro) energy conversion, and ocean thermal energy conversions; and, finally, utilization via organic materials.

Probably the best prospects for the early utilization of solar energy are the *heating and cooling of buildings*. Detailed analyses have shown that the economic potential for systems that combine solar heating and cooling is much better than for single-purpose systems. The degree of development of solar water heating, space heating, and space cooling systems differs

[11]The comments on fusion are adapted from the Stockholm International Peace Research Institute report *Nuclear Age*, SIPRI, Stockholm, 1975, pp. 31-32; and "Alternative Energy Sources for the US", op. cit., pp. 12-13.

[12]"Alternative Energy Sources for the US", op. cit., pp. 13 et seq.

markedly. While solar water heaters are commercially available, solar space heating technology is considerably less developed, as is solar energy use for air conditioning. Various assessments of the potential application of solar systems for buildings in the US have been made. These studies conclude that the potential market is large. Of the projected 60 million new buildings which will be constructed during the next 25 years, almost two-thirds could be candidates for solar heating and cooling systems.[13]

About two percent of total US energy consumption for the year 2000 could be supplied by solar heating and cooling of buildings. A figure of less than one percent is probably a more likely expectation. Process-heat applications could perhaps account for a similar percentage of the US energy consumption.[13]

The use of solar energy to generate electricity on a large scale is also worthy of serious consideration. The major obstacles to full implementation of these schemes are the high cost of all the designs under consideration and the fear that large scale solar plant construction might entail excessive land requirements.

A photovoltaic solar plant would probably place the greatest demands on land usage. For the photovoltaic system, the area requirements for the collector are very large. Assuming a 10 percent efficiency for the solar cells, a 100 MWe solar plant would require a collector area of about 21 square miles.

To make extensive application economically feasible, the cost of solar arrays would have to be reduced by a factor of a hundred or more from present levels. Simultaneously, the efficiencies of solar cells will have to be improved and their effective lifetimes extended. It is estimated that 1 to 3 percent of US energy consumption might be supplied by photovoltaic systems in the year 2000.

Wind energy conversion systems are another possibility. The maximum total energy that could be extracted from the significant wind potential for generating electric power is low. While there is no major technical obstacle to the implementation of windmill systems, the problem of finding acceptable sites remains. A 1000 MWe power plant might require as many as fifty 20 MWe aeroturbine units and occupy an area of perhaps 50 square miles.

Another energy extraction scheme centers on the utilization of *ocean thermal gradients*. While ocean surface temperatures in the tropical and subtropical regions are in the mid-eighties, the temperatures drop to the mid-thirties a few thousand feet below the surface. The resulting 50°F temperature difference can be used to run a low-efficiency heat engine. Usable sites obviously are limited and transmission costs would be high. The technical problems of massive underwater construction—coping with corrosion and biofouling—are considerable.

The last major category of solar energy applications concerns the *utilization of solar energy via organic materials*. The sources of these organic materials are the managed production of plant tissue and the collection and utilization of available organic wastes. The utilization of existing organic

[13]"Alternative Energy Sources for the United States", op. cit., pp. 13 et seq.

wastes represents a sizeable potential energy source. The total amount of organic waste produced in the US is more than 2 billion tons annually, and at least 880 million tons of this waste is comprised of moisture- and ash-free organics. While the problems of widespread utilization of organic materials to produce energy are manifold, the motivation for utilizing the organic materials available in our wastes is high.

To sum up, the overall prospects for limited utilization in this century of solar energy seem good. Potential solar energy resources that could be tapped are not insignificant. Reasonably realistic implementation which may be achievable could result in our obtaining 10 percent of our energy needs from the solar source by the year 2000. It is clear however, that there will be no substantial contribution prior to 1990 and that solar power does not represent a viable alternative in the near or intermediate term.

III. ESTABLISHED COMMERCIAL TECHNOLOGY FOR NUCLEAR STEAM SUPPLY SYSTEMS IN USE TODAY

A. An Historical Perspective

The concept that useful energy could be extracted from the atom began with the discovery of radioactivity in the early 1900's. The discovery of fission in 1939 and subsequent development of potential chain reaction models culminated with the achievement of the first chain reacting system in 1942[14]. Nuclear development efforts then went forward in parallel towards the construction of nuclear weapons and nuclear reactors for the production of plutonium and plants to produce highly enriched uranium. Initial operation of the production reactors at Hanford, Washington, occurred in 1944. These initial production reactors were light water cooled and graphite moderated. Additional production reactors constructed at the Savannah River plant in South Carolina were heavy water cooled and heavy water moderated.

Following World War II, the nuclear industry moved forward from this relatively small base. A number of reactor concepts were developed and evaluated throughout the world. In the US, the selection of the pressurized water reactor for utilization in the Naval reactor program gave an added impetus for the utilization of water reactor systems. With the commissioning of the Nautilus submarine in 1955 and subsequent construction and operation of the nuclear plant at Shippingport, Pennsylvania, in 1957, the light water reactor industry was born in the US. The development of the boiling water reactor (BWR) in 1954, and ultimate utilization in the Dresden, Illinois, plant in 1959, established this reactor type as the other major contributor to the US nuclear power program. As the advantages of nuclear power came to be recognized, utility commitments were made for this power source. With the awarding to General Electric of the General Public Utilities' Oyster Creek, New Jersey, contract in 1963, nuclear power was accepted as a commercial reality in the US.

The current US commitment to light water reactor systems is large. By 1985, an estimated 185,000 megawatts of electrical generation capacity in the form of light water reactors may be in place. The magnitude of this commitment and the time required to bring about any change in basic technology indicate that the light water reactor is the reactor type we must concern ourselves with when evaluating nuclear fuel supply questions in the US.

Different priorities and interests in other nations dictated the development of different reactor systems. Because UK production reactors were gas

[14] The power output of this 'pile' although not converted into useful energy was initially one-half watt and finally 200 watts.

cooled, early UK nuclear power plants were primarily gas cooled, as were those developed by France. Canada chose the heavy water, natural uranium, pressure tube system.

B. How a Nuclear Power Plant Works

The production of electricity in a nuclear power plant is similar in principle to the production of electricity in any large power station. Steam which is heated in a boiler is released to a turbine generator where the heat energy contained in the steam is converted to electricity. In a nuclear power plant, the heat source is nuclear fuel instead of coal, oil or gas. Instead of the heat resulting from combustion, a chemical reaction, the heat produced in the nuclear fuel results from fission, a physical reaction.

Despite numerous possible variations in the design and components of nuclear reactor systems, there are a number of general features which all reactors possess. The reactor consists of an active core in which the fission reaction is sustained and in which most of the energy is released as heat. The core contains the fissile fuel material, the moderator which slows down the neutrons released by fission so they can be used efficiently, and the coolant which removes the heat produced during fission. Reactors are generally classified by the coolant used, the moderator used and the speed at which neutrons move in the core (e.g., fast or at slower thermal speed). In some cases, the coolant also serves as a moderator. The coolant is circulated through the core to absorb heat and then used either directly or with a secondary steam supply system to generate electricity in a conventional manner by means of turbine generators.

C. The Light Water Reactor

In light water reactors (LWRs), "light" or ordinary water acts as both the moderator and the coolant. The principal consequences of the use of light water as a moderator is the requirement to use lightly (2 to 4% U^{235}) enriched uranium as a fuel, thus necessitating the availability of uranium enrichment facilities.[15]

There are two types of light water reactors in wide use: boiling water reactors (BWRs) where the coolant is allowed to boil and thus can be used directly in the turbine generators; and pressurized water reactors (PWRs) where the heat is transferred by the coolant to a secondary system for the generation of steam.

The light water reactor is the work horse of today's nuclear electric power generation. In 1973, light water reactors accounted for 52 percent of the total operating power reactors in the world (75 out of 144). In the same year, LWRs accounted for 73 percent of the world's total nuclear electric power generation capacity (34,200 megawatts out of the world total of 46,670 megawatts). Of the 75 light water power reactors operating in 1973, 38 were installed in the US; 7 each in the USSR, West Germany and Japan; 3 in Switzerland; 2 each in East Germany, India, Italy, the Netherlands and

[15]See Appendix B.II., "A Layman's Guide to the Nuclear Fuel Cycle", for details.

Spain; and one each in Belgium, France and Sweden. Almost all of the major components of these LWRs were manufactured by or under license from the US, with the exception of those in the USSR and East Germany.

With West Germany, France, Sweden and Japan, as well as the USSR, today capable of exporting LWRs, and another five or six countries soon to be, LWRs will continue to hold the major share of the world reactor market. US construction of LWRs for domestic purposes will probably account for about 50 percent of world sales.

D. Heavy Water Reactor (CANDU)

The heavy water reactor is moderated with water enriched in deuterium, or heavy hydrogen, instead of normal hydrogen. The reactor coolant is in pressure tubes and a secondary system is used for the production of light water steam for turbine generators, as in the pressurized water reactor. The use of heavy water as a moderator means that natural uranium may be used as a fuel, thus eliminating the need for the enrichment step in the nuclear fuel cycle (which is necessary in the case of the light water reactor). The system does however require the production of heavy water in substantial quantities. This results in a demand for enriched water analogous to that for enriched uranium in the light water reactor. Moreover, the problem of safeguarding heavy water reactors is affected by the frequency of refueling as well as the flexibility of producing plutonium in this system.

The most advanced heavy water system is the CANDU reactor, manufactured by Canada. The CANDU is being vigorously promoted and is expected to make a modest contribution to energy generation in the near and intermediate term. Heavy water systems accounted for approximately 7 percent of total installed nuclear capacity worldwide at the end of 1973.

E. The Gas Cooled Reactor

The *high temperature* gas cooled reactor, and its advanced concept, the gas cooled breeder reactor, are prospective technologies, and are treated in the following chapter. However, *low temperature* gas cooled (i.e., Magnox) reactors are an established technology now in use for some nuclear steam supply systems. At the end of 1973, gas cooled reactor systems accounted for approximately 20 percent of total world nuclear power plants. This share is expected to decrease substantially as LWR-based nuclear power programs continue to expand on a worldwide basis.

* * *

The systems reviewed in this chapter are those which have evolved from the several reactor concepts developed in the early 1950's, and are established nuclear power technology in commercial use today. There are, however, a number of new systems, and improvements to existing systems, which could be developed. Cost improvements in existing reactor systems may be obtained by "closing the fuel cycle," that is by the recycling after fuel discharge of uranium remaining after, and plutonium generated during, power production. (The advantages and disadvantages of recycling are discussed in detail in Chapter V.) Cost reductions in the LWR fuel cycle may also be obtained from the development of improved materials allowing higher temperatures and therefore more efficient operation of turbine generators; from the development of more effective management techniques; and from other improvements.

IV. PROSPECTIVE TECHNOLOGY FOR FUTURE NUCLEAR STEAM SUPPLY SYSTEMS

The major changes over the long term in the future development of nuclear power are likely to be those evolving from the use of prospective reactor systems: liquid metal fast breeder reactors; high temperature gas cooled reactors, and (possibly) light water breeder reactors. These are discussed below.

A. Liquid Metal Fast Breeder Reactors (LMFBR)

The liquid metal fast breeder reactor (LMFBR) is a system characterized by fast (i.e., unmoderated) neutrons and the use of molten sodium as a coolant. It is designed so that in addition to the power producing core common to other reactor types, there is a surrounding blanket of uranium in which neutrons from the core are captured and plutonium is produced. Hence the name "breeder", a plant which over a ten to twenty year period of time produces more fuel than it burns.

The characteristics of a breeder reactor require that the fuel contain more fissile material than the "thermal" reactors discussed earlier (i.e., LWR, GCR and CANDU). Fast reactors require 15 percent plutonium in uranium for efficient operation.

The development of the liquid metal fast breeder reactor began in the US and, following World War II, except for the operation of production reactors, fast reactor technology moved in parallel with thermal reactor development. It is interesting to note that the first electricity generated from a nuclear plant was from a fast breeder, Experimental Breeder Reactor I (EBR-I) at the Idaho nuclear testing station, on December 20, 1951.

From EBR-I, the US went on to build EBR-II, a 20 MWe plant commissioned in Idaho in 1965. At the same time, in Europe, the demonstration fast reactor at Douneray was being developed by the UK (commissioned 1963) and the Rhapsodie reactor by France (commissioned 1967). From this comparable base, the breeder programs in the US and the rest of the world have proceeded at different paces.

In parallel with the successful development of the EBR-II plant, the Fermi plant—a 60 MWe breeder—was developed and constructed outside of Detroit. Construction delays and other difficulties plagued the completion of the plant. Fermi was finally commissioned and operated until 1968 when a flow blockage in the core caused the plant to shut down. Fermi was fully repaired after the accident and operated briefly, but was subsequently shut down because its metal fuel core did not provide useful information for today's ceramic fuel concept.

The breeder program in the US suffered continued difficulties during the middle sixties. The construction of the Fast Flux Test Facility (FFTF), originally scheduled to be completed in 1974, was delayed and the plant is now scheduled for operation in 1978. Difficulties in both the breeder development program and the management of committed breeder projects, which by the early 1970s included a Fast Breeder Demonstration Plant, caused the US program to be substantially delayed. Operation of the demonstration plant is now scheduled for 1983. Consequently, the commercial development of a US breeder is not expected to occur until the mid 1990s, given the time needed to demonstrate the technology, economics, regulations and safeguards which are prerequisites to such commercialization.

While the US was involved in difficulties in its breeder program, European nations were moving steadily ahead. Demonstration plants of the 200-300 MWe class were authorized by the British, the French, the Germans and the Russians. These plants are currently in operation (France and USSR) or scheduled for operation in the near future (UK and Germany). The USSR has also placed a 600 MWe demonstration plant in operation. There are currently plans by France and the UK to place 1000 MWe units in service before the mid-eighties. Thus, there are prospects for commercialization of the breeder reactor in Europe at a much earlier date than in the US. (France is currently projecting reliance on the breeder of up to 30-40 percent of installed nuclear generating capacity by the year 2000.)

The advantages of the liquid metal fast breeder reactor fall into two categories: advantages associated with the heat transfer characteristics of the liquid metal system, and advantages associated with fuel cycle costs and the utilization of fuel resources. The liquid metal system is advantageous because it produces higher temperature, higher quality steam. This results in greater efficiency in the production of electricity, thus yielding lower costs and more complete use of resources. The higher efficiency also results in less waste heat being produced which must subsequently be disposed of. In this respect, it can be argued that the LMFBR affects the thermal environment less then the light water reactors, and, in fact, is comparable in this respect to present day fossil fuel plants.

The second and perhaps more significant set of advantages resulting from breeder utilization are those associated with resource use and fuel cycle costs. Because the LMFBR produces fuel that can be subsequently re-used, the amount of energy that can be extracted from existing uranium resources is increased by a factor of 70.

The disadvantages of the breeder, apart from unresolved technical problems, relate to the development of public policy regarding the protection of the public health and to the question of public security. The development of breeder reactors on a large scale raises the same safeguard questions as the recycling of plutonium. These questions are examined in some detail in Chapter V.

B. Light Water Breeder Reactors (LWBR)

The light water breeder reactor (LWBR) uses the uranium-thorium fuel cycle and light water as a coolant. The LWBR was rejected as a primary choice for a breeder system in the early 1960s when it was evaluated relative to the LMFBR as a commercial breeder because of its relatively small breeding gain and anticipated materials problems. However, LWBR development has continued on a modest scale, and a demonstration core is scheduled for insertion in the Shippingport reactor in 1976.

The system offers the advantage of the LMFBR in better utilization of fuel resources (although the improvements in resource utilization over LWR systems are not nearly as dramatic as with the LMFBR). The ability to use existing technology developed from LWR systems is a prospective benefit for the LWBR. Because of the ability to use existing LWR technology such as pumps, turbine generators, etc., there is a chance that when the technology is proven[16], and should the economics appear favorable, a reasonably rapid introduction of the LWBR could be possible. The demonstration program which the LWBR will undergo should be watched closely so that, if successful, an early assessment of its introduction on a wide scale can be made.

C. High Temperature Gas Cooled Reactors (HTGR)

The high temperature gas cooled reactor (HTGR) and its advanced concept, the gas cooled breeder reactor (GCBR) use the uranium-thorium fuel cycle in contrast to the uranium-plutonium fuel cycle used by light water reactors and liquid metal fast breeder reactors[17]. Additionally, the HTGR is graphite moderated.

The advantages associated with the HTGR involve the possibility of both reduced cost of power and better utilization of fuel resources. Another advantage of the HTGR is that it can serve industrial purposes other than power generation, i.e., process heat, a development currently being explored in Germany. However, it is unlikely that these systems will see any early significant utilization in the US in the light of recent commercial events. The HTGR was being promoted by a single supplier, Gulf-Shell General Atomics (GA). In the course of marketing, GA sold nine large HTGR systems to five utilities. Recently, eight orders were cancelled. Subsequently, GA announced major layoffs and has essentially withdrawn from the market. The sole remaining commercial HTGR system, the Fort St.

---

[16] 1979 is the time for first discharge from the Shippingport Atomic Power Station of a LWBR fuel charge. This will provide for a careful experimentally based assessment of the technological merit and prospective economics.

[17] A footnote on two fuel cycles:

There are two fuel cycles potentially of use in power reactors:

1. The uranium-plutonium fuel cycle which is used in LWR and LMFBR systems, and
2. The uranium-thorium fuel cycle which has potential use in the HTGR and LWBR.

In the uranium-plutonium fuel cycle, uranium 235 is the initial fissile material. Plutonium 239 is also bred in the uranium 238 blanket initially present, and then subsequently used as a fissile material:

$$U^{238} + n \rightarrow Pu^{239}$$

In the uranium-thorium fuel cycle, uranium 235 is the initial fissile material. Uranium 233 is bred in the thorium 232 blanket initially present and then subsequently used as a fissile material:

$$Th^{232} + n \rightarrow Th^{233}\ \beta- \rightarrow Pu^{233}\ \beta- \rightarrow U^{233}$$

Vrain, Colorado, 330 MWe electric plant, originally scheduled for operation in 1972, has been plagued with startup difficulties and is not expected to operate until mid-1976.

V. NUCLEAR FUEL FOR LIGHT WATER REACTORS

A. Nuclear Fuel—what is it?

Today natural uranium and enriched uranium are the principal nuclear fuels in wide use, as they have been for the last quarter century. Other nuclear fuels such as thorium, recycled uranium, and plutonium may be of importance in the future as they become commercially available.

In addition to the US, Canada, France (primarily in Gabon and Niger) and South Africa are the current major non-Communist producers of uranium. As well, Australia has major proven reserves of uranium which are expected to come into production in the near future. Mining uranium ore is the first in a series of steps known as the nuclear fuel cycle, set out in *Figure V.1*. (A detailed description and historical perspective of the nuclear fuel cycle, the dynamics of the nuclear fuel supply equation, nuclear fuel cycle investment and economics are treated in *Appendix B*.)

The mined ore is sent to a mill where uranium concentrate is produced. (Uranium concentrate is often referred to as "yellowcake" and has the chemical symbol U_3O_8.) There are fourteen mills presently producing uranium concentrate in the US. The uranium concentrate is further refined to uranium oxide for use in the (Canadian) CANDU deuterium reactors; other reactors, and in particular light water reactors, require that the uranium be *enriched* before it is used as fuel.

Simply stated, *uranium enrichment* is the process by which natural uranium is physically altered into a richer mixture of the fissile isotope U^{235}, which can then be used as fuel in nuclear power plants to produce electricity.

B. How is nuclear fuel produced?

Natural uranium contains only seven-tenths of one percent of the energy-producing U^{235} isotope of uranium. The remainder of the natural uranium, U^{238}, is the non-fissile part. The enrichment process increases the U^{235} content to the two to four percent level needed to fuel light water nuclear reactors.

This is accomplished by first converting the uranium concentrate or yellowcake into uranium hexafluoride, or UF_6. Uranium hexafluoride is a gas at conditions near room temperature and pressure. Two UF_6 conversion plants are operating in the US, and one each in Canada, the UK and France. UF_6 conversion is also carried out in the Soviet Union.

At the present time, there are three uranium enrichment processes under consideration for commercial use in the US: the established gaseous diffusion process used in ERDA's three plants; the gaseous centrifuge process; and the laser isotope separation process. France, the UK and the USSR currently use the gaseous diffusion process for uranium enrichment. In gaseous diffusion, the uranium hexafluoride gas is pumped through a

Figure V.1

The Light Water Reactor Nuclear Fuel Cycle

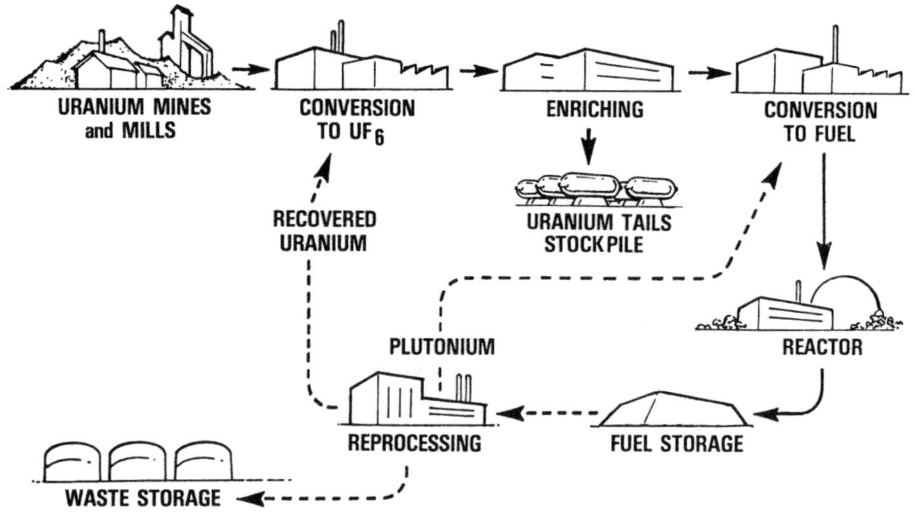

semipermeable membrane which acts as a filter. The lighter U^{235} uranium isotope filters through the membrane more readily than the heavier U^{238} isotope, thus collecting into a stream with a higher-than-normal concentration of U^{235}. The centrifuge process utilizes centrifugal force to separate the heavier from the lighter uranium isotopes, thus increasing the concentration of U^{235}. In the laser isotope separation process, which is in the early stage of development, U^{235} atoms or molecules are selectively ionized to provide for their concentration through conventional chemical separation techniques.

Because uranium enriched to very high levels can be used in nuclear weapons, uranium enrichment technology has been and remains classified. The classification applies to the technology and economics of the process and to the design and manufacturing of some of the equipment used.

Uranium enrichment services are sold in units of separative work (SWU), which is a measure of the amount of effort required to separate U^{235} from U^{238}. The capacity of a uranium enrichment facility is therefore measured in separative work units (SWU's).

The uranium enrichment plant owner (today exclusively the Government in the US) processes customer-owned uranium in his enrichment plant. The plant owner does not sell enriched uranium as such; he sells the service of enriching the customer's uranium. Once the desired enrichment is completed, the material is shipped to a fuel fabrication plant. There the enriched uranium is converted to uranium dioxide (UO_2), formed into pellets and placed in zirconium tubes. The tubes are assembled into bundles, called "fuel assemblies", and sent to nuclear power plants. Seven US companies are involved in the fabrication of nuclear fuel.

C. How is nuclear fuel used?

The fuel assembly is inserted into the power reactor where the fuel is used to generate heat through the process of fission, as described in *Appendix B*. From this point on, the system follows the conventional steps of using the heat to produce steam, which in turn drives steam turbines that turn electric generators. Nuclear fuel is thus one primary source of energy used to produce a secondary source of energy, electricity, that can be easily delivered to the consumer and turned off and on at full power, instantly, conveniently, and cleanly at the point of use.

D. What happens to nuclear fuel after it is used?

After the fuel is used in the nuclear power plant, it is discharged and cooled in a large water basin at the plant. Most spent fuel is currently being stored on reactor sites. The spent fuel could in the future then be sent to a chemical reprocessing plant (none is in commercial operation in the US today). There, the uranium and reactor-produced plutonium would be separated from the highly radioactive waste products generated as a result of fission in the nuclear power reactor. The radioactive wastes, converted into a solid, would then be shipped to a Government repository. The recovered uranium would be converted again into the hexafluoride gas and reinserted into the enrichment plant for re-enrichment.

The extracted plutonium, which is also a fissile material, could be used as fuel in nuclear power plants. If use of plutonium is authorized in the US by the US Nuclear Regulatory Commission, plutonium oxide would be sent to a fuel fabrication plant where it is mixed with uranium oxide and formed into pellets for nuclear fuel. This process is known as *plutonium* recycle.

E. What unusual problems are involved in the nuclear fuel cycle?

The major unresolved issues and constraints associated with the demand, supply and use of nuclear fuel will be identified and examined in the next three chapters. Here, however, it is useful to set out the several developments and primary technical problems intrinsic to the nuclear fuel cycle itself.

First, it is necessary to assess the adequacy of the resource base for nuclear power: the supply of uranium available to meet demand. While the demand for uranium depends on the demand for nuclear electrical energy, there is a necessary technical refinement in this relationship: demand for uranium depends not only on demand for nuclear power, but also on whether or not the recycling of uranium (and plutonium) occurs, and on how much and what type of *enrichment* capacity is available.

The supply of reprocessed fuel will affect the demand for fresh uranium. The recycling of both uranium and plutonium could reduce uranium resource requirements by as much as 35 percent.

The effective utilization of the enrichment process and the process type also affect the demand for uranium. The utilization of enrichment capacity is measured by a "tails assay", i.e., how much U^{235} is left in the depleted portion of the uranium feed. Operation of enrichment facilities at high tails assay decreases the amount of enrichment services required but requires more uranium. Conversely, low tails assay means less uranium is used but that more enrichment services are required. (See *Appendix B* for details.) Tails assay variations in gaseous diffusion and centrifuge plants can affect uranium demand by as much as 40 percent.

The unique potential of laser enrichment to separate a significant fraction of U^{235} from U^{238} could result in the ability to economically maintain tail assays at a relatively low level (e.g., 0.05%). The impact of laser enrichment on uranium demand derives from its extraction of more U^{235} from natural uranium feed than the gaseous diffusion or the centrifuge enrichment process. Savings of up to 40 percent in uranium feed are projected. For this reason, laser development, now in the early stages, is being accelerated by industry and the Government in the hope of early commercialization.

Consequently, the uranium demand and supply data discussed in the following chapters are projected in the light of five factors:

(i) nuclear electrical energy demand;
(ii) The required *level* of enrichment for light water reactors, 2 to 4% U^{235};
(iii) the *amount* of enrichment (measured by assaying the leftover tails) that will result in the most economic choice between using more uranium feed vs. more enrichment to reach the required 2 to 4% U^{235} level;
(iv) The *type* of enrichment process used: gaseous diffusion, centrifuge or laser;
(v) the effect of recycling or no recycling on uranium and enrichment services demand.

Another technical aspect of the nuclear fuel cycle that presents unusual problems concerns fuel storage, the reprocessing of spent fuel, and the disposal of radioactive wastes. (See *Appendix B*).

The decision on whether or not to reprocess spent fuel depends upon whether or not it is advantageous because recycled fuel is economic; or because use of a finite and depleting uranium resource base needs to be prolonged; or both. The advantages and disadvantages of recycling are examined in the following section of this chapter.

Spent fuel discharged from reactors must be stored in large pools in order to cool. If the cooled spent fuel is *not* ultimately reprocessed, the spent fuel must be permanently disposed of as high level waste.

On the other hand, if reprocessing does take place eventually, as is likely, at least two and possibly three significant products would result:

- first, the unused *uranium* in the spent fuel would be extracted chemically and processed for re-enrichment or direct re-use in reactors;
- second, the *plutonium* produced in the reactors and contained in the spent fuel would be separated from the uranium and the radioactive

waste, and subsequently mixed with uranium and fabricated into new fuel for use in reactors.

- finally, there would be residual radioactive wastes which would be shipped to a Government repository for permanent secure storage.

Consequently, there are two key decisions involved in what happens to spent fuel:

(1) whether or not to reprocess spent fuel in order to recycle;

(2) how to dispose of the spent fuel or radioactive waste in a permanent and safe way.

Reprocessing spent fuel in order to recycle uranium, while leaving the plutonium in the radioactive waste and storing it retrievably for possible future use, is technologically feasible. Because the economics of recycling uranium alone are not considered attractive, in the US or abroad, it is not deemed realistic at this time to plan on uranium recycle alone. However, recycling uranium alone, without recovering plutonium, could possibly be attractive from the point of view of safeguards implementation and cost.

A second and perhaps more attractive plutonium strategy is the dilution of plutonium in uranium fuel. The reprocessor would dilute uranium to various levels dependent upon the amount of plutonium available and the plutonium/uranium "blend" that can be used in reactors of a given design.

The difficulty with the second option, however, is that in order to blend the plutonium with uranium to form new fuel, the plutonium must first be separated out chemically from the spent fuel residue. In other words, at one point in reprocessing, prior to recycling, the plutonium exists in a purified form, suitable for use in nuclear explosive devices, before being mixed with uranium. And at that point theft or diversion of weapons grade plutonium would theoretically be possible. The longer the period of time the plutonium is accessible in purified form, the greater the burden on safeguards.

In order that reprocessing and recycle of uranium and plutonium may be compared with the alternative of continuing to store spent fuel, advantages and disadvantages of recycle are next examined.

F. Advantages and Disadvantages of Uranium and Plutonium Recycle

The advantages of recycling uranium and plutonium accrue in two areas. The primary advantage would be a reduction in need for both uranium resources and enrichment services. Possible savings of up to 35 percent in uranium supplies and up to 20 percent in enrichment capacity are estimated by most analysts. The fact that these residual materials represent a known "mineable" resource should be a substantial incentive for investment in its recovery. Fuel cycle cost savings associated with full recycle (i.e., recycling both uranium and plutonium) are estimated to be from 5 to 20 percent depending upon the impact which recycling may have upon the cost of newly mined uranium. While the fuel cycle capital investment savings are offset by required investment in reprocessing plants, this reprocessing base can be ultimately utilized as a base for breeder reactor fuel supply.

The disadvantages of recycling plutonium and uranium are argued on the grounds of security as well as health, safety, and environmental considerations. The recycle and potential diversion of plutonium raises serious questions. A program for recycling would require the construction of additional reprocessing plants. The co-location of such plants with existing and planned fuel fabrication facilities may alleviate the problem of transporting divertable materials.

The decision to proceed with the recycling of uranium and plutonium should be dependent upon the balancing of the positive impact of additional resources on the extent and economics of nuclear power against the negative impact of increased risk of diversion and the attendant development of opportunities for the proliferation of plutonium. Reprocessing and recycling may be optional. What is mandatory is an adequate and economic supply of fuel to meet demand. An increment of that supply may be obtained through recycling.

France, Germany, and the UK apparently see recycling as a necessity, given their projections of uranium supply and enrichment capacity, and, in France, given the current emphasis placed on introducing breeder reactors. Japan, among others, may agree. The US is proceeding to demonstrate the technology, economics, safeguards and other regulation associated with reprocessing in order to proceed with later industrial operations as resource and economic considerations dictate.

VI. DEMAND AND SUPPLY OF NUCLEAR FUEL FOR LIGHT WATER REACTORS

The parameters of future total energy demand, electrical energy demand and nuclear electrical energy demand for the US and the rest of the world were examined in Chapter I. Actual trends as well as future commitments indicating increased reliance on nuclear power were observed both here and abroad. While relatively low fuel cost experience and projections were seen to favor nuclear power from light water reactors, in comparison with the alternatives examined in Chapters II, III and IV, a number of unresolved problems or uncertainties affect supply of nuclear fuel needed to meet projected nuclear electrical energy demand.

In order to develop a clearer picture of the present state of the art of nuclear energy, the future supply of nuclear fuel will be examined in the context of the specific constraints and unresolved issues associated with its demand, supply and use. To do so, it is useful to begin by quantifying the current and forecast demand for nuclear fuel.

A. How much is required?

The demand for nuclear fuel is determined by the estimated growth of nuclear power, which was seen in Chapter I to be in the range of 14-16 percent annually in the US and the industrialized countries of Western Europe, Canada and Japan. This overall demand for nuclear fuel can be separated into demand for components of the nuclear fuel cycle described in Chapter V:

- — natural uranium (U_3O_8)
- — conversion to uranium hexafluoride (UF_6)
- — uranium enrichment services
- — fuel element fabrication
- — spent fuel storage
- — reprocessing of spent fuel
- — recycle of recovered uranium and plutonium
- — radioactive waste management.

US and foreign current and projected demand (corresponding to the moderate nuclear power growth case given in Table I.4) is discussed next for each of these components of the nuclear fuel cycle:

Natural Uranium Demand

The projected US, foreign non-Communist and Communist annual and cumulative natural uranium demand through the end of the century is given in *Tables VI.1* and *VI.2* respectively. Table VI.1 forecasts that US annual demand in the years 1980, 1985 and 2000 will amount to 2½, 5 and 10-13 times the 1974 domestic production of 12,600 short tons of U_3O_8. Table VI.2 estimates that cumulative US demand will exceed the firm reserves *plus* probable additional resources by the mid-1990's.

Since nuclear reactor commitments may require the determination of an adequate supply base prior to plant commitment, the projections (in Table VI.2) must be examined in the light of the need for a future reserve. In other words, because it takes 7-10 years between exploration for uranium and bringing a mine-mill facility into actual operation, the cumulative demand in Table VI.2 may be regarded as shifting 7 to 10 years closer to 1975. For example, if a 10 year lead time is assumed, then the 1980 ten-year forward demand for about 700,000 to 800,000 short tons of U_3O_8 would require *doubling* current reserves under $15 in the next five years if higher cost uranium is to be avoided. A decision to stockpile requirements in advance would of course further increase forward demand.

The uranium resource significance of recycling may be seen in Table VI.2. No recycling in the US results in an additional cumulative demand of 27,000 short tons of U_3O_8 in 1985 (i.e., more than twice current annual production capacity) and about 277,000 short tons in 1995. Corresponding increases in cumulative demand of the foreign non-Communist nations of the world in the case of no recycle are 61,000 short tons and about 300,000 short tons for 1985 and 1995 respectively.

Recycling (beginning in 1979) could reduce cumulative uranium demand in the US by about 7 percent for 1985 and 20 percent for 1995; corresponding foreign non-Communist demand would be reduced by about 12 percent for 1985 and about 15 percent for 1995.

Demand for Conversion to Uranium Hexafluoride, UF_6

The projected US annual demand for conversion of natural uranium to uranium hexafluoride through the year 2000 is given in *Figure VI.1*. The data are given in tons of uranium as uranium hexafluoride, for the case of recycle and the case of no-recycle. In both cases, the demand is approximately 20,000 tons in 1980, and diverges moderately thereafter.

Uranium Enrichment Demand

Projected world annual and cumulative demand for uranium enrichment separative work (SWUs), assuming availability of uranium, through the year 2000, is given in *Tables VI.4* and *VI.5*. Projected world cumulative demand without recycling amounts to a total just over 100 million SWUs in 1980. The data are given for recycle as well as no-recycle, for the US, foreign non-Communist and Communist nations, but do not include government (i.e., military weapons) requirements, or national or plant stockpile provisions, or enrichment plant working inventories.

Fuel Fabrication Demand

The projected US nuclear power annual demand for enriched uranium dioxide (UO_2) fuel *fabrication* is given in *Figure VI.2* in metric tons of heavy metal, amounting to almost 4,000 tons in 1980.

Spent Fuel Storage Demand

Demand for US light water reactor spent fuel storage is projected in metric tons of heavy metal in *Table VI.6* and in approximate number of fuel assembly units in *Figure VI.3*. The spent and discharged fuel must be stored for six months or more of cooling prior to transportation to a reprocessing plant for possible recovery of uranium and extraction of plutonium, which could in turn be recycled. Spent fuel storage demand depends upon the decision on when to reprocess spent fuel, as illustrated in *Figure VI.4*.

Reprocessing Demand for Light Water Reactors

The projected US nuclear power annual and cumulative demand for reprocessing are given in *Table VI.7* through the year 2000. The US cumulative demand is projected as 6,695 metric tons of heavy metal (MTHM) in 1980; 20,639 MTHM in 1985; and 165,990 MTHM by the year 2000.

B. What are current and planned nuclear fuel supplies, and are there future supply deficiencies?

The nuclear fuel supply capacity necessary to meet the world demand for nuclear fuel projected in the preceding section is quantified below, with a view to forecasting future supply deficiencies. In light of these supply projections, constraints in meeting US and world demand are then identified, together with major unresolved issues in demand, supply and use of nuclear fuel here and abroad.

Natural Uranium Supply

During 1974, sixteen US uranium milling companies represented US operating capacity and produced about 12,600 short tons of U_3O_8, over 95 percent from the 37 largest US mines and less than 5 percent from over twice as many smaller mines. In 1975, the US produced 11,500 short tons of U_3O_8, less than the amount produced in 1974.

The current annual production capacity of the non-Communist nations, and their production capacity of 36,550 short tons U_3O_8 attainable in 1978 are set out in *Table VI.8*. US annual production capacity is seen as capable of expanding to about 24,700 short tons U_3O_8 by the middle of 1978.

US producers' and buyers' inventories today include, in the form of natural uranium, the equivalent of about 25,000 short tons of U_3O_8. The US ERDA has a stockpile of enriched uranium equivalent to about 50,000 tons of natural uranium, and about 30,000 equivalent tons of natural uranium in the form of obsolete weapons: ERDA has indicated that these stockpiles will be reserved for national contingency purposes. Foreign non-Communist stockpiles total about 48,000 short tons, which may or may not be made available to meet world demand.

If the US projected supply is to meet US projected demand, substantial additions to US supply capacity must be in place and operating five years from now, 1981, as it is at that point that US demand as projected above will have surpassed the US capacity estimated for 1978 and existing US non-government stockpiles. If a sizeable national stockpile is deemed prudent, then new supply must be available perhaps one year earlier, 1980. Failure to bring new supply on line will require the import of foreign uranium supply.

The impact of recycling is not seen as altering the 1980 demand and supply situation significantly. US projected annual demand would exceed the currently expandable supply (24,700 short tons) and begin to deplete stockpiles in 1978; and would be 25 percent greater than annual supply by 1980. *Table VI.9* indicates new mines and mills needed by the US.

The projected foreign and non-Communist world supply picture is similar to the US outlook. Non-Communist world capacity is projected at about 55,000 short tons of U_3O_8. Annual demand would exceed annual supply in 1978, but stockpiles would allow for perhaps one more year before new capacity would be needed. Prospective world demand for new supply capacity is summarized in Table VI.9 along with that for the US.

In summary, US natural uranium supply production, if it is to meet prospective demand, should be increased by a factor of about five in the next 10 years and by a factor of about 13 or more in the next 25 years. World supply should be increased by about the same or slightly larger factors in the same time periods.

Conversion Services Supply

The existing and planned capacity of the non-Communist world to convert U_3O_8 to UF_6 is summarized in *Table VI.10*. The current US capacity is 53 percent of world capacity. Planned capacity for 1978 exceeds existing capacity by 26 percent in the US and by 65 percent in the rest of the non-Communist world.

Beginning about 1980, new US conversion capacity will be required to meet US demand (about nine new plants with an individual capacity of 10,000 short tons of U_3O_8 per year (STU/yr) between 1980 and 2000, in the case of recycle; and in the case of no-recycle, about twelve new 10,000 STU/yr plants over the same period of time). If US conversion plants are needed to meet foreign demand, then additional plants over that indicated would be required.

However, as noted above, foreign capacity is already undergoing expansion, and the current policies of the other major uranium resource countries, Canada, South Africa, and Australia, indicate that they plan to build conversion facilities in correspondence with their natural uranium production. US utilities purchasing uranium outside the US may therefore be required by the producer to convert it outside the US, thus reducing US domestic demand for conversion facilities.

Uranium Enrichment Supply

World uranium enrichment supply from plants currently in operation is projected in *Table IV.11*. The table compares the cumulative non-Communist world demand with the cumulative total world supply through 1987, including USSR exports to Western Europe. The table shows that the US (i.e., ERDA) currently produces 93 percent of the non-Communist supply. For the US, projected annual demand (including exports) exceeds the projected supply from existing plants in 1979; cumulative demand is estimated as equal to cumulative supply in 1983.

New supplies currently planned and authorized are given in *Table VI.12*, indicating that ERDA and the French-led consortium "Eurodif", which also includes Italy, Spain, Belgium, and indirectly Iran, represent 91 percent of currently planned and authorized new capacity. Although URENCO, the centrifuge partnership formed by the UK, Germany and the Netherlands, hoped to reach an annual supply level of about 10 million SWU in the mid-1980s, its current commitments and authorization only take it to the 2 million SWU per year level indicated.

Cumulative *firm* world supply is projected in Table VI.12 to fall short of non-Communist world nuclear power demand in 1988. *Table VI.13* shows how the addition of new 6 million SWU per year plants beginning in 1984, and totalling about 20 such plants by the year 2000, would meet the projected non-Communist world demand through the year 2000.

Fuel Fabrication Supply

Current US fuel fabrication capacity is more than 50 percent greater than domestic demand because of foreign market demand. The current capacity is scheduled for an increase of 20 percent by 1978 and projected for an increase of 300 percent by 1980, at which time it would probably represent about two-thirds of world supply.

Spent Fuel Reprocessing Supply

Until reprocessing plants become operational, spent fuel must be stored in existing cooling pools at the reactors, at reprocessing plants, or in new "stand-alone" storage facilities. *Table VI.14* projects the new reprocessing capacity needed to meet anticipated domestic demand for reprocessing and also work off the backlog of spent fuel already accumulating. In addition to the two existing US reprocessing plants (Barnwell and West Valley), which are not yet authorized for commercial operation, additional plants are projected in the table as required, beginning in 1986, to work off the backlog by 1996.

Radioactive Waste Management

The solidification and permanent disposal of *low-level* radioactive wastes is an ongoing and proven industrial process. Technology for the permanent disposal of *high-level* radioactive waste has been developed but has not yet been demonstrated in the US or abroad. High-level wastes produced to date have therefore been stored, more or less successfully, in temporary containments. Because some of these containments have deteriorated with age, some leakage has occurred. To date, no damage has been reported, and ERDA is currently replacing these older containments. But the fact remains that a permanent solution should be demonstrated and implemented.

VII. THE CONSTRAINTS ON MEETING NUCLEAR FUEL DEMAND

In the preceding chapter, current and projected US and world demand for the various components of the nuclear fuel cycle have been compared with existing and planned supply capacities. Supply deficiencies have been identified, perhaps most strikingly in the world supply of natural uranium and the US supply of uranium enrichment services, and also in projected deficiencies in the world supply of reprocessing services and spent fuel storage.

While the specific demand and supply projections advanced here may prove to be more or less accurate over the next 25 years, the trends are observable today, and the deficiencies foreseen are of sufficient magnitude and imminence to warrant active consideration by national supply planners now.

What then are the constraints which supply planners and the supply industry are experiencing in meeting nuclear fuel demand? Are these constraints unique to US experience, or are they shared—or compounded—by other industrialized nations, and the developing world?

The constraints on meeting demand are:

- Physical resources
- Trade policies and technology transfer
- Financial resources
- Financial incentives
- Institutional capacity and responsibility
- Regulation: rates, environmental, health and safety
- Government security classification
- Technology
- Public/private sector decision-making
- Public perceptions and attitudes
- Economics
- Safeguards

These constraints on meeting demand are assessed here in the context of US domestic supply planning and, when appropriate, the interdependent supply planning of the other major current and potential users and suppliers of nuclear fuels.

A. Physical Resources and Supply Capacities

The primary natural resource limitation to nuclear fuel supply is the availability of uranium. In the United States the expansion of uranium mining and milling operations is complicated by the fact that actual uranium resources have not as yet been accurately assessed. Inadequate exploration and delineation of uranium reserves is the result of the historically cyclical demand for uranium, increasing dramatically but temporarily in the 1950s for weapons production, and now again increasing dramatically for power production. Additional complications stem from such problems as the availability of skilled miners and equipment (drill rigs, drag lines, etc.) and the competition with other mining industries—especially coal—for these components of the production process, as well as the usual difficulties associated with expanding the capacity of a resource industry and its supporting infrastructure from a small base in a short time.

A national uranium resource evaluation (NURE) is currently being undertaken by US ERDA, but is not scheduled for completion before 1980. Some encouragement may be taken from the fact that ERDA's preliminary results indicate that additional domestic supplies may be found, although their quality and quantity can not yet be predicted with any high degree of accuracy. Because of uncertain US demand on the world uranium market, Western Europe and Japan may not be able to make realistic supply assumptions.

This supply uncertainty is compounded by the restrictive export policies of foreign uranium producers, in particular Canada and South Africa. If and when the US finds itself (marginally) dependent upon foreign sources for uranium, there is no guarantee of continued US access to that supply in competition with other (e.g., Japanese, West European) buyers. The world uranium market consists of a limited number of producing and exporting countries, whose resource allocation and trade policies today tend to discourage the export of natural uranium. The rationale behind these practices is discussed below (under "Trade Policies").

In additional to the potential shortfall of US and world uranium supply, and the question of world market availability, the secondary constraints on meeting demand for nuclear fuel are the projected insufficiencies of US and world enrichment and reprocessing capacity. Lack of reprocessing capacity is not an immediate constraint because temporary spent fuel storage is an available alternative. Lack of reprocessing capacity does result, however, in an increase in the demand for newly mined uranium and enrichment services. As detailed earlier, most countries which have ongoing programs for the development and use of nuclear electric power have selected the light water reactor as the energy source for meeting their near-term requirements, the significant exceptions being Canada and the UK. As light water reactors require enriched uranium fuel, there is the dual question of how to provide enough uranium ore *and* enrichment capacity to fuel the numbers of reactors planned. Moreover, in order to reduce the demand for natural uranium, reprocessing and recycling of spent fuel may prove to be desirable.

Conversion of uranium to uranium hexafluoride, the fabrication and storage of fuel elements, and low-level waste disposal are all commercially available, proven, industrial-type technologies. In Europe, both centrifuge enrichment and fuel reprocessing technologies are today in industrial operation: three centrifuge enrichment plants, one in the UK and two in the Netherlands, form part of the European consortium URENCO; and two reprocessing plants, in France and the UK, are nearing completion for reprocessing spent fuel from light water reactors.

As the US and the rest of the world increase reliance on nuclear energy, expansion of US enrichment capacity is a necessity. The entire capacity of the three US (Government) enrichment plants, including that being derived from an expansion program, has already been fully committed (indeed overcommitted) under long-term contracts with domestic and foreign utilities. Since it is expected to take seven to eight years to provide large new plants, commitments to expand US enrichment capacity must be made without further delay if important adverse effects on US nuclear fuels policy are to be avoided.

In expanding US enrichment capacity a major question exists as to whether the government or private industry will supply the next and future increments of enrichment services. While the policy of the current Administration is to transfer this responsibility from the government to the private sector, it is not yet clear that the transition will occur in time to meet demand.

Enrichment services are in fact a type of industrial activity usually found in the private sector of the economy; and the private sector has been urged to enter the enrichment business. Private industry has so far not been able to elicit the large amounts of long-term capital required to do so (discussed below), with the result that interested firms have concluded that some form of government financial assurances would be necessary to effect the transition of enrichment services to a private industry.

In the case of reprocessing, the situation is somewhat different but the result is identical. Significant private capital investment has already been made in developing US reprocessing capacity, and private sector pilot plants have been built. The technology and economics of reprocessing, however, remain to be proven in the context of yet uncertain regulation, and consequently the private sector is looking to the Government to assume the additional and considerable financial risk of demonstrating the technology on a viable commercial scale.

B. Trade Policies

Other than the US, only eight countries are known to have major deposits of natural uranium: Australia, Canada, France, Gabon, Niger, South Africa, Sweden and the Soviet Union. And, in addition to the US, only five nations or groups of nations are currently seeking to provide the export of uranium enrichment *services*: Eurodif (the French-led international consortium including Spain, Italy, Belgium and Iran); URENCO (the consortium of West Germany, the Netherlands and the UK); West Germany; South Africa; and the Soviet Union. World nuclear fuel trade policies and practices are summarized country-by-country, in regard to the export of uranium and of enrichment services, in *Appendix C*. The limited number of national sources of uranium, and of sources of enrichment services, may be contrasted with the many nations indicated in Chapter I as having nuclear power plants in operation, under construction or planned.

In September 1974, the Canadian Government announced a new uranium policy with two primary objectives:

(1) to insure at least a 30-year reserve of nuclear fuel for all existing, committed and planned reactors in Canada for any ten-year forward period;
(2) to ensure that sufficient uranium production capacity is available for the Canadian domestic nuclear power program to reach its full potential.

It is estimated that over and above the export commitments and domestic allocations, almost 50 percent of the Canadian uranium resources are uncommitted for future export or domestic needs. The present expectation is that between now and 1985 there can be firm export commitments, between 1985 and 1990 conditional export commitments, and after 1990 no exports because of domestic requirements.

The restrictive trade policies and practices of South Africa concerning uranium export have been designed to encourage the development or acquisition of domestic uranium conversion and enrichment technology in order to have a secure and independent domestic supply of energy, and, at least equally important, to enhance the value of uranium prior to export. The industrial technology for conversion (i.e., for converting uranium to a gaseous fluoride compound) is not difficult to develop or acquire; but the highly complex technology needed to enrich uranium is.

US enrichment technology is classified by the US Government and in practice has not been exported. Because nuclear weapons can be made from very highly enriched uranium, or from plutonium obtained through reprocessing spent fuel, the US has consistently opposed the export of both enrichment and reprocessing technologies. Enrichment and reprocessing technology are included in the recently announced intent of West Germany to sell to Brazil complete nuclear fuel cycle capability.

The US over the years has taken the initiative in working with other concerned countries to insure that civil nuclear exports were used only for peaceful purposes. Recently, the US has had a number of bilateral and multilateral discussions with other exporters of nuclear equipment and technology with a view to devising common rules of the road concerning application of safeguards and related controls. As a result of such discussions, the US has decided to apply certain principles to its future nuclear exports. Most of these are consistent with current US practice; some are new. All are designed to further non-proliferation goals while facilitating the sale and export of nuclear equipment, materials and technology necessary to meet the world's growing energy needs.

While the results of the negotiations have not been reported officially, recent unofficial reports indicate that agreement has been reached on the safeguards and physical protection related to proliferation that would be required in the recipient country as a condition of supply in connection with such sales in the future. Despite the ban on US sales of enrichment and reprocessing technology and equipment, safeguarded sales from other countries will therefore presumably continue. Further safeguards may result from the proposed creation of multinational, perhaps regional nuclear fuel supply centers, a development supported by the US Secretary of State before the United Nations in September, 1975,[17] and under active study in the International Atomic Energy Agency. In addition to taking advantage of potential economies of scale and enhancing supply security, such centers would internationalize the ownership, control and operation of enrichment and reprocessing technologies and fuel fabricating, making diversion or theft for the manufacture of weapons less likely than would dispersed national facilities. The US Government is considering permitting the transfer of US enrichment and reprocessing technology to such multinational centers. Various countries want a nuclear fuel supply industry of their own today for reasons of energy diversification and assurance, and in some cases for reasons of prestige or for the option of developing weapons capability. Whether they would consider that multinational, regional centers meet their interests will depend in part upon the likely efficiency and reliability of such centers, and whether or not they are able to find alternative sources for supply of the requisite technology.

C. Financial Resources

The capital investment required for nuclear fuel supply and services represents approximately 10 percent of the capital investment in a nuclear power plant. The viability of the investment in a nuclear power plant depends completely upon assurance of its fuel supply.

While the capital investment in nuclear *power plants* has traditionally been organized by and through the utility industry, the capital investment in nuclear *fuel supply* has not: the nuclear fuel supply industry is new, non-integrated, largely derivative from other non-nuclear fuel supply industries, and therefore lacks an historical investment trackrecord.

The financial resource constraints on meeting nuclear fuel demand are two: the magnitude of the capital required to finance the production of adequate nuclear fuel supplies; and the financial risks involved for investors. Major uncertainties include not only the future demand for nuclear fuels but also the important lack of experience in measuring costs and profits of commercial enrichment and reprocessing plants.

Specific uncertainties concerning the capital requirements for expanding the (private) uranium mining and milling industry derive from the difficulty in extrapolating cost experience to the grades of ore which will have to be mined in the future. Also, the capital requirement for spent fuel storage and waste disposal is in part contingent upon the unresolved issues of reprocessing and recycling.

Subject to these and other uncertainties, projections obtained from the Edison Electric Institute's recent analysis[18] of the demand for each compo-

---

[17] Multinational regional nuclear fuel supply centers were first proposed in 1968 by the British Pugwash movement, but at the time failed to elicit widespread interest.

[18] See footnote (5), page 5.

Table VII.1

APPROXIMATE NUCLEAR FUEL CYCLE CAPITAL REQUIRED PER 1,000 MWe OF NUCLEAR PLANT WITH Pu RECYCLE (million 1975 $)

Mining and Milling	10.8
Conversion	0.9
Enrichment	28.0
Fabrication	3.1
Reprocessing	14.0
Waste Facilities	2.8
Total	59.6

Source:
EEI (Edison Electric Institute Nuclear Fuels Supply Study Program, 1976)

nent of the nuclear fuel cycle indicate that capital investment on the order of $15 to 19 billion would be required in the US through 1985, and roughly $60 billion through the year 2000. In estimating the *net new investment needed*, allowance should be made for supply industry currently in operation and worth about $8 billion, making the required new investment through 1985 on the order of $7 to 11 billion ($15 to 19 billion less the $8 billion worth of existing supply industry).

Table VII.1 presents the approximate capital investment required for the various components of the nuclear fuel cycle per 1,000 megawatts of nuclear generating capacity. The data correspond to that given in *Table VII.2* for 120,000 MWe of capacity.

The major question surrounding financial requirements relates to the possibility of attracting sufficient investment in a timely manner. Enrichment plants, for example, require large investments and, in order to be economic, high utilization. They therefore become a risk by virtue of the magnitude of the investment required to build plants of an economic size. In enrichment on an economic scale, the increments in which capital must be put together for gaseous diffusion plants are $3 billion for each 9×10^6 SWU capacity (the economic optimum plant size). Centrifuge and laser enrichment plants can be constructed in smaller option units, i.e., 3×10^6 SWU. Thus the investment required for a centrifuge plant of 3×10^6 SWU is $1 billion. This may be decreased in that these concepts are hoped to be less capital intensive than the diffusion alternative.

D. Financial Incentives

The kinds of question the investment community and the corporate board room ask are: Is the resource base of economically competitive uranium likely to be adequate to support the estimated requirements? Are the licensing and regulatory requirements well enough established to support capital and operating cost estimates and lead to a responsible basis for pricing? Are

Table VII.2

APPROXIMATE NUMBER OF MAJOR FUEL CYCLE SUPPLY PLANTS AND APPROXIMATE CAPITAL INVESTMENT TO SUPPORT ~120 × 1,000 MWe OF LWR NUCLEAR POWER FOR FULL RECYCLE AND NO RECYCLE (a)

	Unit Annual Capacity	Capacity Units (b)	Unit Cost Million $	Pu and U Recycle: No. of Plants	Pu and U Recycle: Capital Cost (c) Billion $	No Recycle: No. of Plants	No Recycle: Capital Cost (c) Billion $
1. Mining-Milling	1,000	STU_3O_8	70 (d)	18.6	1.30 (d)	25.80	1.81 (d)
2. Conversion	5,000	MTU	40	2.86	0.11	3.96	0.16
3. Enrichment	9	Million SWU	3,500	0.96	3.36	1.28	4.50
4. Fabrication							
Uranium	1,500	MTHM	110	2.04	0.22	2.72	0.30
MOX	200	MTHM	30	5.10	0.15	—	—
5. Reprocessing	1,500	MTHM	700	2.4	1.68	—	—
6. Waste Treatment	1,500	MTHM	100	2.4	0.24	—	—
7. Waste Storage and Disposal	12,000	Ft^3	150	0.6	0.09	— (e)	— (e)
8. Interim Spent Fuel Storage	2,000	MTHM	36	—	— (f)	15	0.54
				Total:	7.15 (d)		7.31 (d)

(a) Assumes 0.3% enrichment plant operating tails assay.
(b) Units as follows:
STU_3O_8 = Short tons natural uranium—yellowcake
MTU = Metric tonnes uranium
SWU = Separative work units
MTHM = Metric tonnes heavy metal (U and Pu)
(c) 1975-dollar approximate estimates.
(d) Dependent upon ore grades mined in the future.
(e) No comparable estimate available. According the total column in the 'No Recycle' columns is an incomplete statement of total costs.
(f) No estimate available. However, about 0.1 or 0.2 investment is already committed.

Note: Not directly useful for 'recycle versus no recycle cost-benefit analyses' since lead and lag times, operating cost, etc., must be considered.
Source: EEI (Edison Electric Institute Nuclear Fuels Supply Study Program, 1976)

the prognoses for local moratoriums on nuclear energy or other extended delays such as to discourage major capital investments at this time? In the final analysis, will the private sector be permitted or encouraged to develop an enrichment industry? Will the venture, given all the uncertainties, be a profitable one? These and similar considerations are real questions and concerns which cannot be dismissed. They must be dealt with effectively or the unavailability of capital could preclude the private sector from developing an adequate nuclear fuel supply industry.

Even when these uncertainties are resolved, there remains the additional constraint imposed on financial incentives by the long lead times required to bring components of the nuclear fuel cycle on stream: 10 years for uranium mining and milling (from exploration to full production); 4 years for conversion plants; 8 years for enrichment facilities; 5 years for fabrication plants; 10 years for reprocessing facilities. The lead times are not, of course, sequential; but they are interactive. Investment in some phases tends to be retarded in the absence of reasonable certainty that other phases will be completed on time.

The continuing evolution of technology in the nuclear field enhances the incentive to favor more rapid turnaround projects. At the level of risk involved, current interest rates are not good indicators of the discounting required to consider the relative profitability of investments five or ten years hence. In effect, the further in the future the return on an investment occurs, the higher the discount rate must be to cover the possibility that more competitive technologies will be developed in the interim.

E. Institutional Capacity and Responsibility

The Administration has proposed the "Nuclear Fuel Assurance Act" to provide government financial guarantees and thereby assure the capital required to build large new increments of enrichment capacity. The proposal was made because of the apparent inability of the investment community to undertake the risks involved in financing new enrichment capacity, and because of the capital formation problems of the electric utility industry. A third reason for this Administration initiative, and perhaps the most important one in the long run, is the growing conviction in the private sector, as well as in the Executive Branch, that the US public and US customers abroad would both be better served by getting the Government out of, and the private sector into, the business of uranium enrichment services—as the preferred way of making that industry more responsive to domestic and world markets.

This transition to private industry in a market economy is thought to be an effective way of eventually permitting the actual economic and social costs of the nuclear energy option to be passed on directly to, and therefore accurately measured by, the consumer. The economics of nuclear energy are advantageous relative to alternative sources of electrical energy, given the increased costs of all fossil fuels resulting in large part from the upward adjustment of oil prices. The "privatization" of enrichment would probably enhance the attractiveness of the nuclear energy option as perceived by the investment community. A number of additional reasons for proceeding today with the creation of a nuclear fuel supply industry have been advanced:

- Uranium enrichment services are now essentially an industrial type of activity. It is not one that can be performed well only by the government.
- Private industry wants to enter the uranium enrichment business.
- Private participation would create a more diverse fuel supply base to support a growing nuclear power industry.
- Private competition should provide incentives, over the long term, for lower costs, improved efficiencies, and technological advancement.

Responding to the Administration's proposal encouraging private entry into the uranium enrichment industry, several industrial firms have made substantial progress in preparing to build, own, and operate enrichment facilities. One venture, that of Uranium Enrichment Associates, has reached the stage where it has proposed construction of a specific plant, is accepting letters of intent from customers for enrichment services, and has taken options on land and electrical power. Other ventures have been organized and plans have been proposed for plants using centrifuge technology to provide the next increments of enrichment capacity, and laser technology to provide eventual additional increments.

It should be kept in mind that the Nuclear Fuel Assurance Bill addresses *only the problem of enrichment supply*. It will not alleviate the financial risk or alter the capital demand of the other components of the nuclear fuel cycle, in particular uranium supply. Expanding enrichment capacity requires large increments of investment capital. Expanding uranium reserves and production capacity, as projected in this study, will require an amount of capital equal to or greater than enrichment, though in more and smaller increments more readily handled in the routine operations of the capital markets. Consequently, with the unique exception of potential government financial guarantees for expanding enrichment capacity, the responsibility for financing and providing nuclear fuel supply today rests with the private sector.

F. Regulations: rate, environmental, health and safety

Regulation is a general constraint imposed by law on the electric utility industry and on its suppliers of nuclear fuels. The major regulations which apply to the electric utility industry concern prices or rates, and environmental, health and safety regulations. Rate regulation is for the most part a state responsibility.

The electric utilities recently have had a capital formation problem in financing expansion of and changeover to nuclear generating capacity, and in financing or acquiring any significant portion of the nuclear fuel supply industry. The reason often given for this is that the investment market has not been favorable to the utilities, primarily because of their low return on investment. Low returns are usually attributed to inflation, increased prices for fuel, and added cost for pollution control equipment and other environmental, health and safety regulations, all of which are not yet fully covered by authorized increases in the rates for electricity paid by consumers. While significant rate increases have been granted, there is a lag between the time when increased costs are incurred and the time when commensurate increases in rates are granted by the regulatory agencies.

The regulation of nuclear power plants and fuel cycle facilities for health and safety and environmental matters is performed for the most part at the Federal level. Constant evolution of these regulations is a major source of uncertainty in the siting, design, construction, operation and economics of nuclear power plants and associated fuel cycle facilities. These changing regulatory requirements have led to a hesitancy in new commitments in these areas. These actual problems in the regulatory process are compounded by a lack of credibility of the regulatory agencies as perceived by some sectors of the public.

G. Government Security Classification

Because US enrichment technology is both Government owned and classified, private industry is having problems assessing the commercial viability and profit potential of going into the enrichment business. The security classification of enrichment technology is imposed by the Government to prevent the diversion of such technology to the production of materials for nuclear weapons, and to prevent its proprietary loss.

The Government has indicated its willingness to share the technology with US firms interested in entering the enrichment business, as well as with foreign nations with which the US may participate in multinational ventures. Despite the fact that the US does not have a monopoly on enrichment technology, it has so far refused to transfer or sell it to other countries on a bilateral basis. As a consequence, the US is today unable to compete in international sales of enrichment technology, as exemplified by the recently announced sale by West Germany to Brazil.

H. Technology

The development of technology has not proved to be a significant constraint on meeting nuclear fuel demand.

I. Public/Private Sector Decision-Making

The role of the Government, and what it will and will not do to support a nuclear fuel industry, remains today the key uncertainty in assuring adequate nuclear fuel supply. As a result, it is difficult to assess what the private sector will do, much less what it could do unaided (or unhindered), as long as its capability and will are masked by the uncertainty regarding Government involvement.

The cumulative uncertainty of pending Government decisions on enrichment and reprocessing is compounded by speculation on the likelihood and level of Federal involvement. Recently the tendency in each instance associated with nuclear fuel supply has been for the Government to propose to intervene: directly, in the case of reprocessing; through elaborate financial guarantees, in the case of "privatization" of enrichment; and through regulation, in the banning of enrichment technology exports and uranium imports. The resulting mix of private and public sector decisions-to-be-made has become so complicated that whatever its outcome, the consequent action may prove to be too late to permit the US to exercise its energy options effectively.

Many of the difficulties facing the energy industries, including the nuclear energy industry, turn out to be largely questions of the present decision-making system. And the competing demands on the capital market over the next decade or so will not encourage energy investment unless some of the risks of decisions in this field are reduced. Many of the risks are man-made, linked to unpredictability of policy, inadequate coordination, ideological swings in Government thinking about its role in energy, Congressional and Executive divisions, and sheer lack of planning.

Within the government, a real coordination system is needed which can provide overall energy policy guidance for specific decisions on issues ranging from rate regulation to environmental control, from enrichment policy to international safeguards policy. This probably requires, to be effective, a public/private advisory system of substantial magnitude which provides advice to the government on the full range of energy issues, and which functions as an educational process for the capital markets.

J. Public Perceptions and Attitudes

To date, the general public has not viewed the immediate problems of the nuclear fuel supply industry in any integrated manner. There has been public awareness of the need to accelerate the development of available domestic energy sources, including nuclear power, to meet energy demand while reducing dependence on imported fuels. But nuclear fuel supply realities have tended to be overshadowed by public interest in short-term issues, such as gasoline lines or prices, and long-term energy supply possibilities, such as solar power, as well as environmental, health and safety concerns. Also, a vocal anti-nuclear movement has been directing small but highly visible and sometimes effective attacks on local governments and utilities, with the consequent possibility of local moratoriums on the installation of additional nuclear generating capacity.

K. Economic Constraints

There are a number of economic constraints common to all current alternative sources of energy, such as the difficulties that arise in predicting growth in total energy and electrical energy demand, or assessing the impact of new environmental, health and safety regulations, or forecasting how much of total energy production will depend upon which alternative. However, there also have been several uncommon factors that pertain to nuclear power growth: high capital costs, long licensing procedures, and long construction times in addition to public acceptance problems. At least they did pertain primarily to nuclear power prior to 1973. More recently, however, these factors are increasingly shared to a significant degree by the other current domestic energy alternatives: oil, gas and coal.

The abrupt adjustment of oil prices established a new plateau for economic assessment of alternatives, but the role of market prices is not clear today. The long term relationship among the alternatives will not become evident for several years. In the meantime, multiple prices can coexist within the marketplace partly because of government regulations, price controls, policies, and commitments already made.

The price of uranium has risen rapidly in recent years. Assessments of world reserves will always be somewhat unreliable. Exploration and identification of new reserves in most non-renewable resources tend to be stepped up during times of economic expansion, sometimes globally and sometimes in connection with the economic growth of particular countries.

L. Safeguards

Nuclear weapons can be manufactured from two products: plutonium and highly enriched uranium. "Safeguards", as used here, are the measures taken to deter, prevent, and detect the proliferation of nuclear weapons and the capability to produce them. Safeguards are therefore applied in three areas: to nuclear *materials*; to the *facilities* which use or process these materials (reprocessing and enrichment capacity); and for the use of *technology* needed to build these supply capacities (reprocessing and enrichment knowhow).

Four kinds of safeguards are applied nationally, by government action and regulation, and internationally, by treaty and intergovernmental agreement, to assure that these materials, facilities and technologies are not diverted to weapons manufacture:

(1) *Physical security*: to prevent the diversion of materials and facilities to the manufacture of weapons.

(2) *Accountability:* to deter diversion of materials by the threat of timely detection.

(3) *Inspection*: to deter diversion of facilities as well as materials by the threat of prompt detection.

(4) *Codes of conduct* (treaties and international agreements): to regulate the international transfer of enrichment and reprocessing technologies, facilities and materials, ensuring that such transfers will be made only if adequate safeguards are in place to deter, prevent and detect diversion.

Whatever the US decision may be concerning the future export of the US enrichment technology, other countries are exporting that technology, and it should therefore be safeguarded. Whatever the US decision may be concerning domestic reprocessing, reprocessing is taking place elsewhere, and should therefore be safeguarded. Consequently, the US is currently seeking further international agreement on safeguards for these two technologies, for their application in production facilities, and for the resulting special nuclear materials that are produced.

Even if the US were to reject further expansion of its domestic nuclear energy industry, Europe, Canada, Japan and many of the developing countries have already committed themselves irrevocably to large scale and widespread nuclear electric power generation. Foreign demand for US nuclear fuel supplies therefore continues and is increasing. This is most evident in the case of US enrichment services, which today contribute 92 percent of the non-Communist countries' supply.

The effects on the US of nuclear energy development in the rest of the world cannot be altered by any US decision regarding the domestic use of nuclear energy. These effects concern the security of the US in a world in which nuclear weapons capability may proliferate. Hence, regardless of whether there is to be stagnation or expansion in the domestic nuclear industry of the US, the US must continue to foster world safeguards in order to deter, prevent and detect national or sub-national nuclear diversion.

In recent Congressional testimony, four basic categories of nuclear weapons proliferation risk were identified[19]:

(1) There is an apparent certainty of continued vertical proliferation, i.e., the US-USSR arms race. Despite the worthwhile effort that has been devoted to the SALT talks, and the modest progress that has been made, the arms race continues. As long as it does, the apparent political logic of horizontal nuclear weapons proliferation becomes extremely difficult to refute.

(2) There is a risk, some would say a certainty, that additional governments may acquire nuclear weapons capabilities. The number of nations possessing nuclear weapons or explosives has gradually increased to at least six: the US, the USSR, Great Britain, France, China, India—and, probably, Israel. The first five countries acquired the fissile materials required for their weapons program from nuclear fuel cycle facilities originally built for military purposes. India is the first country to have exploded a nuclear device using materials produced and processed in a facility originally intended as a research reactor.

(3) There is a risk that non-governmental groups may acquire nuclear weapons from military stocks, or explosive-grade fissile materials from military channels or such materials from civilian nuclear power programs. Stolen fissile materials might be fabricated into explosives or poisons, used for extortion of the enterprise or government stolen from, or sold on a black market. The motives for non-governmental nuclear weapon proliferation include political terror and monetary profit. The groups that might engage in this type of proliferation include an individual acting alone, a terrorist organization, or a criminal group organized for financial gain.

---

[19]Testimony given before the U.S. Senate Committee on Government Operations, by Professor Mason Willrich, January 19, 1976. The following analysis of safeguards is taken from this testimony.

(4) There may be a risk of nuclear weapon proliferation resulting from political revolutions or realignments in countries where nuclear weapons are deployed or where explosive-grade fissile materials are present.

There is a variety of safeguard measures designed to reduce various risks of nuclear weapon proliferation. Although the various proliferation risks are interrelated, the focus here is on safeguards applicable to the nuclear fuel cycle.

Existing safeguards measures to reduce the risk of governmental proliferation include:

The Non-Proliferation Treaty (NPT)

The NPT embodies an obligation on the part of the signatory non-nuclear-weapon parties not to acquire nuclear weapons or explosive devices. The NPT thus contains a self-denying ordinance that is binding under international law. Any party may, however, withdraw if it believes its supreme interests are jeopardized. The Treaty obligations may operate as a useful constraint in complex domestic decision-making processes, as well as provide a legitimate external source of leverage on those processes.

In addition, the NPT requires non-nuclear-weapon parties to accept International Atomic Energy Agency (IAEA) safeguards on all their peaceful nuclear activities, and requires all NPT parties to require IAEA safeguards on their nuclear exports to non-nuclear-weapon countries, whether or not the importer is an NPT party. Thus exports to nuclear-weapon states need not be safeguarded. Moreover, a non-nuclear-weapon state not a party to the NPT may continue to receive imports from NPT parties, subject to safeguards, while still maintaining other nuclear activities in its territory free of safeguards (unlike a non-nuclear-weapon party to the NPT).

IAEA Safeguards

Safeguards administered by the International Atomic Energy Agency (IAEA) are intended to *verify* that nuclear materials subject to safeguards are not diverted to any proscribed purpose. These safeguards are aimed at national governments. Verification may serve as a deterrent to diversion in some situations, but IAEA safeguards are not designed to actually prevent diversion. The main safeguards measure is materials accountancy. Important corollary measures are containment (e.g., seals) and surveillance (e.g., automatic cameras). The efficacy of international safeguards rests heavily on the stringency of materials accountancy systems. Given the size of material flows likely to build up within nuclear power fuel cycles in the future, materials accountancy may not be sufficient as a safeguards measure against governmental diversion. Increased reliance will therefore have to be made on containment and surveillance, internationally, and physical security, domestically.

The IAEA system offers a useful opportunity for national governments to reassure the world community that they are not in the process of acquiring nuclear weapons. However, if a government decides to divert nuclear material subject to safeguards for use in a weapons program, the most that should be expected from the IAEA system is a yellow warning light as a result of frustration of the inspection process.

National Safeguards

Safeguards administered by national governments and applicable to fuel cycle facilities and materials at such facilities and in transit are intended primarily to *prevent* unlawful diversion and secondarily to detect any unlawful diversion that might have occurred. National safeguards are now being strengthened because the risk of nuclear theft has increased concomitant with the worldwide expansion of commercial nuclear power.

Many of the key decisions concerning national safeguards in the US and abroad to deal with the risk of non-governmental nuclear weapon proliferation remain to be made. US national safeguards have been substantially strengthened since the end of 1973. But decisions remain pending on:

- the co-location of key fuel cycle facilities in order to eliminate potentially vulnerable transportation links;
- the creation of a federal nuclear materials security force;
- further improvement of physical protection and accountability standards; and
- most importantly, the terms and conditions of plutonium recycle in the LWR fuel cycle.

VIII. THE MAJOR UNRESOLVED ISSUES AND OPTIONS IN DEMAND, SUPPLY AND USE OF NUCLEAR FUELS FOR LIGHT WATER REACTORS

In the preceding chapter the general constraints to meeting demand for nuclear fuels for light water reactors were examined. Consideration of these constraints leads to selection of certain unresolved major issues which are identified and further discussed in this chapter.

These issues are:

- Governmental and public perceptions of the adequacy of the resource base
- Public/private sector responsibilities in the US
- Reprocessing and recycle of plutonium and uranium
- Physical security and US domestic safeguards—with or without plutonium
- Radioactive waste management
- Import/export policies
- International safeguards: accountability; physical security; constraints on technology transfers.

A. Government and Public Perceptions of the Adequacy of the Uranium Resource Base

The likelihood of additional US uranium resources has been indicated recently by the preliminary results of the US Government's national uranium resource evaluation. Consequently, the problem inherent in the lack of a complete assessment of total US natural uranium reserves is not the imminent depletion of those resources. Instead, because the national uranium resource evaluation is not scheduled to be completed until 1980, the immediate problem is that of continued public and governmental concern as to the adequacy of the resource base and the correctness of commitments to additional nuclear power plants.

An additional problem is how to transfer potential uranium finds into actual uranium reserves. This can be accomplished only by mounting a timely national uranium exploration and reserve development program, and by the management of the timely production of the uranium as a vital and perhaps rate limiting step following resource evaluation and the translation of potential uranium supplies into actual uranium reserves. How timely such a program turns out to be depends upon market perceptions of potential profitability and the development of a uranium production industry on a national scale capable of meeting projected demand.

Through the Organization of Nuclear Energy Producers (OPEN), some European utilities are today in the process of creating a buffer to potential shortfalls of uranium supply. An attempt is being made to alleviate individual scarcity through industry-wide sharing, by pooling uranium stockpiles currently held by utilities, and periodically adjusting delivery schedules and allocating new uranium supplies. The pooling of uranium stockpiles in the US might prove useful as a buffer for individual US utilities. It would not, however, alleviate the resource, reserve development and production problems noted above.

In part because of the above, and in part to avail itself of economic supply from world resources, the US should consider importing more uranium supplies. Current US regulation does not permit importation of foreign uranium today. Modest quantities can be imported beginning in 1977, and increasing thereafter, with no restriction on imports as of 1984. However, this system of authorizing uranium imports in incremental steps is probably incompatible with the current demand and supply realities. Opening the door to unlimited uranium imports now would help meet current and future demand at economic prices. But US reserves and production capacity must be expanded if the US is not to repeat with uranium the mistake it made with oil: overdependence on foreign supply.

Of the potential non-US sources of natural uranium supply, US interest will in the future tend to focus on Australia. France, which has the highest European demand for uranium, traditionally processes, converts and enriches the supplies of ore from its former colonies, Gabon and Niger. South Africa has its own ambitious enrichment program designed to enhance the value of its uranium exports. Sweden's uranium resources, while significant, can only be developed at a cost sufficiently high as to be prohibitive to date. The Soviet Union, which like the US does not yet know accurately the extent of its uranium resources, has so far refused to export uranium yellowcake. In Canada, as indicated earlier, uranium resources are reserved first for Canadian use: Canada's new energy policy requires that sufficient uranium reserves and production capacity be maintained to meet a projected 30-year nuclear fuel demand for all existing, committed and planned reactors in Canada over any ten-year forward period.

Consequently, Australia is the most recently developing foreign source of abundant uranium available to the US and promises to assume large importance in US energy planning. Recent indications (detailed in *Appendix C*) confirm that Australia is now prepared to resume contracting for the export of large quantities of uranium beginning in 1979. The problem is to organize or at least to evolve an orderly and effective world uranium market.

B. Public/Private Sector Responsibilities in the US

Two general conclusions which relate to private and public sector responsibilities have emerged in the course of the Working Group's deliberations and are discussed here. First, all of the social (environmental, health and safety) and economic costs of alternative—and additional—energy supplies should be expressed insofar as possible directly in the price of the product to the final consumer, in order that clear and informed choices may be made between alternatives and for additional increments of supply. Distortion of actual costs must therefore be minimized, with the definitive dollar vote to be made efficiently and directly in the marketplace to the maximum extent possible.

Second, what the private sector and the marketplace clearly can *not* do, must be either accomplished by the government and the private sector in tandem, or rejected by both together, in the full knowledge of all of the social and economic benefits and costs involved.

What these two general conclusions mean in practice is that private sector capability and will should be the alternative of first resort, and government responsibility the alternative of last resort. Between these two categoric alternatives is a spectrum of increasing government intervention in the otherwise normal functions of the market place. By consistently choosing the least government financial intervention, distortion of actual economic and social costs and benefits, as perceived by the consumer, can be minimized.

There are, however, functions which are clearly beyond the responsibility and/or the capability of the private sector: the international and domestic safeguarding of the technology; capacity and material for producing nuclear weapons; the secure storage and safe disposal of highly radioactive nuclear wastes; and the international coordination of nuclear trade policies and practices, consistent with desired levels of economic growth and agreed measures to prevent nuclear proliferation.

Beyond these assigned responsibilities, there are grey areas where the need for and extent of future government involvement has not been clear. For example, in the past, financing the expansion of enrichment supply capacity has been the responsibility of the government, as that fuel cycle component was originally developed and owned by the government. As a consequence, the actual social and economic costs of producing nuclear electrical energy using this government-owned supply capacity have been partly masked.

By getting the private sector into the business of enrichment, projected nuclear fuel demand can be met, and actual costs can be demonstrated to the advantage of the nuclear option and its attractiveness to the consumer here and abroad. The necessary capability currently exists in the private sector to take over from government the responsibility for building new enrichment capacity, using proven gaseous diffusion enrichment technology, and eventually to take over the existing government-owned enrichment capacity. Because of its role in developing the technology, the government will receive royalties on its use by the private sector, as well as taxes on the profits generated. Similarly, when the government-developed centrifuge and laser enrichment technologies are commercialized by the private sector, royalties and taxes will flow to the government. Moreover, if the government succeeds in demonstrating the technology and economics of reprocessing, the private sector may well be willing to make the investment required to create the necessary reprocessing and recycling industry.

The Administration agrees that the "privatization" of enrichment services today, and of reprocessing services tomorrow, will enhance the supply security and economics of nuclear power. However, the problem is that the private capital needed to do the job has not been elicited from the investment community. And this in turn is because the private capital market has not yet perceived nuclear fuel industry potential returns as outweighing possible risks. As a result, the companies interested in getting into the enrichment business have concluded that government financial guarantees are needed now in order to accomplish the transition in a timely fashion.

There is a clear advantage in having the private sector rather than the government responsible for the nuclear fuel supply capacity needed to meet US and foreign demand: the incentives and competition result over time in lower costs, improved efficiencies, technological innovation, and a diversity of supply that enhances independent supply security and expands the national tax base. The problem therefore is how to make the private sector function more effectively rather than how to alleviate problems of shortage and credibility of continuing supply to be provided through increased government financial intervention.

C. Reprocessing and Recycle of Plutonium and Uranium

Reprocessing and recycle programs are being actively pursued in many countries outside the US. Thus, even in the absence of activities in these areas in the US, there is likely to be a viable governmentally owned or controlled reprocessing and recycle industry in the rest of the world. The rationale for these programs varies, including concerns about the availability of uranium resources, improvement in fuel cycle economics and fueling of prospective breeder programs.

In 1970, the United Kingdom and France undertook an initiative to demonstrate that between the aggressive entry of private industry in the United States into the reprocessing of irradiated fuel, and the existing reprocessing capacity of the UK and France which had been built for military purposes, there was at that time a substantial overcapacity of reprocessing facilities in the non-Communist world. Shortly thereafter United Reprocessors Ltd. (Unirep) was formed, a French/British/German joint venture promising to reprocess, on a doorstep delivery and pickup basis, any and all irradiated fuels produced anywhere in Europe or, for that matter, in any non-Communist countries (as Japan had expressed early interest).

Twelve German utilities in the summer of 1975 created PWK (Planning Company for the Reprocessing of Nuclear Fuel). This company proposes to finance and build the first commercial nuclear fuel reprocessing plant in Germany, a 1500 ton per year facility to go on stream in about 1982.

In Belgium the utilities took the initiative in forming, in concert with the Belgian government, a mixed company called Belgoprocess. Its objective was to purchase the plant of Eurochemic at Mol-Donk and to possibly re-activate it and also construct a new plant of 300-450 tons/year capacity at the same site. The plan has been frustrated by the delays in Belgian government approval for its part in financing as a partner in the plan.

Swiss utilities also appear to regard reprocessing as primarily an affair of the utilities. The utilities seem to recognize and accept that financial participation in some European facility may offer the only realistic expectation of relief from an ever growing inventory of spent fuel.

The Japanese utilities are urgently seeking to expand their own limited reprocessing capacity, preferably not on Japanese soil. One suggestion has been a joint venture with another country able to provide the site for a plant.

For the United States, whether or not plutonium will prove to be a needed and economic secondary source of light water reactor fuel supply will depend upon the availability and cost of fresh uranium and the cost of enrichment services. Plutonium fuel would in any case be required for fast breeder reactors, should that technology become available on a commercial scale. Consequently, once reprocessing is commercially demonstrated, it will be necessary to weigh the marginal advantages of using plutonium in light water reactors, against its eventual use as the initial fuel for breeders.

D. Physical Security and US Domestic Safeguards

Should the US private sector enter the enrichment business, safeguards procedures would have to be in place and demonstrated to be adequate in preventing diversion of the enrichment process or product for weapons use.

With or without plutonium recycle in the US, and indeed whether or not the US proceeds with reprocessing, plutonium will probably be made available as a potential fuel for use abroad because of plans already underway in Western Europe and Japan to utilize reprocessing technology, and to recycle plutonium. Consequently, even if the US decided to forego the benefits of this step in the fuel supply, the US might be affected by its use elsewhere.

E. Radioactive Waste Management

The solidification and permanent disposal of low-level radioactive wastes is an ongoing and proven industrial process. Technology for the permanent disposal of high-level radioactive waste has been developed but has not yet been demonstrated, in the US or abroad. High-level wastes produced to date have therefore been stored, more or less successfully, in temporary containments. Because some of these containments have deteriorated with age, some leakage has occurred. ERDA is currently replacing these older containments. While government and industry continue to assert that the problem is manageable, there is continued public concern as to the long-term manageability of highly radioactive wastes.

F. Import/Export Policies

The US faces today the specific decision whether to permit US exports of enrichment and reprocessing technology and capacity, or to continue to forego such sales and profits in a unilateral effort to limit the possibilities of proliferation, thus leaving the market to others. The export policies of the uranium producing countries, with the exception of the US and Australia, currently tend to discourage the export of natural uranium.

Formation of a formal (OPEC-like) uranium cartel could occur. Reserves are, so far at least, concentrated in only a few countries. However, the market perception may well be that informal action by each exporter will amount to the same thing as formal cartelization. Where numbers of sellers are few, the possibility for tacit cooperation is always great. If those few countries or companies which possess a significant portion of the non-Communist world's economically marketable uranium also control nationally, or as the principal participant in a multinational arrangement, a significant portion of the non-Communist countries' enrichment capacity, then by choosing different tails options (more uranium and less enrichment, or vice versa), these few countries or companies potentially can control the non-Communist world price of uranium and enrichment. This potential development could be effectively acted upon by the US only by the rapid expansion of domestic uranium reserves and production capacity, and enrichment capacity.

G. International Safeguards

The potential proliferation of nuclear weapons is facilitated by the fact that the technologies for enriching uranium and recovering plutonium are being sold and exported today. Even without such sales, it is clearly impossible to stop other countries from developing these technologies indigenously.

To meet this situation, the US has proposed that these two processes be owned, controlled and used primarily in multinational nuclear fuel supply centers. This development could alleviate the demand for additional transfers and development of enrichment and reprocessing capability under national control. However, the would-be importers would have to be satisfied that multinational centers clearly provided a consistently secure source of supply. They would also have to pledge themselves to forego nuclear weapons capability.

The international safeguards system described in Chapter VII currently verifies the international *accountability* for the products of national enrichment and reprocessing plants. The *physical security* of these products and processes could usefully be enhanced through additional international agreement or treaty.

In addition to physical security, a number of potential problem areas should be examined in the hope of developing or supplementing safeguards. These include the following questions:

- If nuclear fuel exports are to be reprocessed, what controls or limitations should the exporting country require of the importer?
- How may trans-shipment of special nuclear materials be safeguarded effectively?
- How may national and international accountability and control systems be improved upon?
- How may the dissemination of technology be controlled?
- Can, or should, so-called peaceful nuclear explosive devices be differentiated from nuclear weapons?
- How may nuclear exports of sensitive technology and capability be constrained?

- How may the IAEA best safeguard the increasing number and sensitivity of facilities that may be assigned to its responsibility?
- How may nuclear exports be controlled among all of the supplier countries[20] with a view to eliminating the possibility of commercial advantage in relaxing safeguards requirements or other anti-proliferation measures?

Effective answers to these questions would in combination reduce but never eliminate the possibilities of proliferation. The task is thus not that of developing an absolute safeguards system. It is rather to make certain that the relative safeguards approaches identified above, together with needed innovations, are implemented in an optimal combination consistent with the assurance of future energy supplies.

The deterrent effect of safeguards, domestically and internationally, is directly related to their credibility. To be credible in the United States today, their implementation must be made known to the public. Moreover, for citizens to choose responsibly more widespread application of nuclear power, they will need to know of all of the measures being taken to protect them.

Over the long term, however, attempts to deal with the proliferation problem through controls on nuclear capabilities can not succeed unless the nuclear supplier countries are able to offer an effective combination of nuclear fuel supply guarantees, national security guarantees—including enhanced conventional military security—and, most important, incentives to forestall independent national development of nuclear weapons capability.

---

[20] Including France and South Africa, which are not party to the NPT, as well as the potential supplier countries, such as Brazil.

IX. POLICY CONCLUSIONS AND RECOMMENDATIONS

The following recommendations are made by the Nuclear Fuels Policy Working Group of the Atlantic Council's Energy Committee:

A. General Conclusions and Recommendations

(1) The utilization of nuclear power is desirable and mandatory in the context of a balanced energy policy leading to increased energy independence by the United States. Nuclear power has been demonstrated to be safe, environmentally acceptable, and economic. The nuclear fuel policy issues addressed below must be resolved on a priority basis such that US and world use of nuclear power may proceed in a timely fashion. Such resolution will allow the orderly transition from an energy supply capability which is based upon depletable resources (oil, gas, coal and uranium) to energy supply capabilities such as fusion, solar energy and the breeder reactor, which are based essentially on non-depletable sources.

(2) The United States should reassert its traditional leadership in international nuclear cooperation by promptly assuring adequate domestic and interdependent nuclear fuel supplies to the non-Communist countries under effective guarantees and safeguards.

(3) The design and implementation of an effective non-proliferation policy should be a high national security goal for the US, today and in the foreseeable future.

(4) US initiatives in international nuclear cooperation should be directed to achieving the following goals:

- assurance of adequate energy supplies for the US and the other non-Communist countries;
- support of the efforts of the nuclear supplier nations and the Parties to the Non-Proliferation Treaty to stem further proliferation of nuclear explosive devices; and
- relief from the pressures on the US and the other non-Communist countries for increasing dependence on imported oil.

B. Recommendations for Prompt Action by the US Government and Private Sector to Assure Domestic Nuclear Fuel Supply

(5) In order to accomplish the goals set out in recommendation #4 above, the US must rapidly expand domestic nuclear fuel supply capacity and attendant technology, as follows:

- increase domestic uranium production capacity;
- expand domestic uranium reserves;
- expand US enrichment capacity;
- demonstrate US reprocessing and recycling technology, economics, and safeguards, as a prerequisite to commercial operation;
- accelerate the development of effective radioactive waste management; and

- accelerate the further development and implementation of safeguards to deter, detect and prevent diversion of nuclear materials, facilities and technology to other than peaceful purposes.

(6) The expansion of domestic nuclear fuel production capacity and technology recommended above should be financed using to the maximum extent the private sources of capital found in domestic and foreign capital markets. Where private capital is not forthcoming, governmental financial guarantees should be considered. This principle should apply to the financing of all components of the nuclear fuel cycle requiring rapid expansion in the US. The principle might usefully be extended to cover capitalization of multinational and bilateral nuclear fuel supply ventures.

(7) The Government should accelerate its current schedule for re-evaluating natural uranium reserves, in order that sound supply planning may proceed.

(8) The Government should promptly remove all restrictions to the import of uranium to the US. Purchases of uranium from all currently producing and potentially exporting countries should be considered. It may also be advisable to consider a uranium inventory policy aimed at reducing possible concerns about over-reliance on foreign uranium sources.

(9) The utility industry, in collaboration with the potential commercial reprocessors, should assure adequate recoverable storage of spent fuel elements until commercial reprocessing becomes available.

(10) In parallel with #9 above, the Government should develop a contingency plan which could result in spent fuel being accepted as the Government's responsibility for further storage and ultimate disposal.

(11) The Government should proceed to cooperate with and encourage industry to demonstrate nuclear fuel reprocessing and recycling technology, economics and safeguards on a commercial scale, in order that decisions as to the further commercial development of reprocessing may be made on a sound and logical basis.

(12) The Government should continue to safely store and should implement plans for the permanent and secure disposal of existing high level radioactive wastes.

(13) The Government should have available an adequate program for the permanent secure disposal of prospective high level radioactive wastes.

C. Recommendations for Prompt Action to Assure International Nuclear Fuel Supply

(14) The US Government and private sector should jointly explore with the traditional and potential foreign buyers of US enrichment services the possibility of exporting US enrichment technology, and eventually reprocessing technology, to multinational and bilateral nuclear fuel supply centers, in order to take full advantage of the prospective economies of scale and supply security such centers would provide, while at the same time safeguarding the further proliferation of enrichment and reprocessing technology.

(15) If and when reprocessing technology is added to multinational and bilateral nuclear fuel supply centers, consideration should be given to co-

locating reprocessing and fuel fabrication facilities at those centers, thereby enhancing international safeguards.

D. Recommendations for Prompt Action to Enhance International Safeguards

(16) The International Atomic Energy Agency (IAEA, Vienna) should be provided adequate funding and manpower, as well as access to national data and facilities, as more and increasingly sensitive facilities become subject to IAEA safeguards.

(17) IAEA accountability should minimize time lags in learning of possible diversion of special nuclear materials.

(18) National verification, accountability, and physical security knowhow should be made available to and pooled through the IAEA.

(19) The IAEA should be provided an appropriate active role in developing safety, environmental and health standards for multinational and bilateral nuclear fuel supply centers.

(20) The IAEA, in cooperation with the other appropriate international organizations, should act as an international clearinghouse for national data on the projected demand for nuclear fuel supply services in all of the developing and the industrialized countries of the world.

(21) The IAEA, in cooperation with the other appropriate international organizations, should act as a clearinghouse for the exchange of information and the exploration of possibilities for sharing technology concerning the various national programs to develop fission breeders and fusion reactors as long-term future energy sources.

(22) Sensitive international nuclear fuel cycle facilities should be colocated to eliminate or reduce vulnerable transportation links.

(23) Safeguards measures to be applied during international transportation of special nuclear materials should be standardized insofar as possible.

(24) International standard measuring systems should be established in order to facilitate the accountability and verification of nuclear materials.

(25) Maximum inspection frequencies and access rights should be required of importers of nuclear materials and knowhow.

(26) The US and other Non-Proliferation Treaty (NPT) proponents should make a determined effort to bring additional nations into the NPT.

(27) An international agreement or treaty should be established to provide for the physical security of nuclear materials and facilities.

(28) Until all supplier nations are party to the NPT, the *ad hoc* and recently enlarged "London Group" of supplier nations should continue to act as the informal bridge to substantive agreement on rigorous safeguards triggered by supplier country exports.

(29) A common policy should be agreed upon by all supplier nations to establish strict standards governing the export of sensitive nuclear materials and technology and to assure their enforcement on a case-by-case basis.

(30) International depositories for plutonium in excess of peaceful national needs—which are provided for in the IAEA Statutes but have not yet been established—should be created now to accept and safely store plutonium under the aegis of IAEA. Such international depositories should include provision for the storage of spent fuel containing plutonium.

APPENDIX A

TABLES AND FIGURES

ENERGY MEASUREMENTS AND CONVERSION FACTORS

Gross Measures of Energy Used in the United States

- The two most commonly used gross measures of US energy supply and demand are quadrillions of Btu (Quads or Btu $\times 10^{15}$) and millions of barrels daily, oil equivalent.
- Two quadrillion Btus per year are roughly equivalent to one million barrels daily (MMB/D).
- By moving straight across *Fig. A.1,* it is possible to match equivalent values in each of the commonly used measures of annual production or distribution for the major fuels and power sources. (see over).
- Uranium is entered as U_3O_8 in units of 10,000 short tons utilized in present thermal reactors.
- Solar energy is entered in units of 1,000 square kilometers of collector surface assuming 50% efficiency and the average amount of sunshine per year in the United States.

Conversion Factors

1 Quad =	180 million barrels of petroleum*
	42 million tons of bituminous coal*
	0.98 trillion cubic feet of natural gas*
	293 billion kilowatt hours of electricity

* These values vary with the quality of fuel actually extracted and represent an average of recent production.

It is also of interest to note that 1 million barrels of oil per day is approximately equal to 2 Quads per year.

Source: ERDA-48

Figure A.1.

Gross Measures of Annual Energy Supply/Demand Commonly Used in the United States

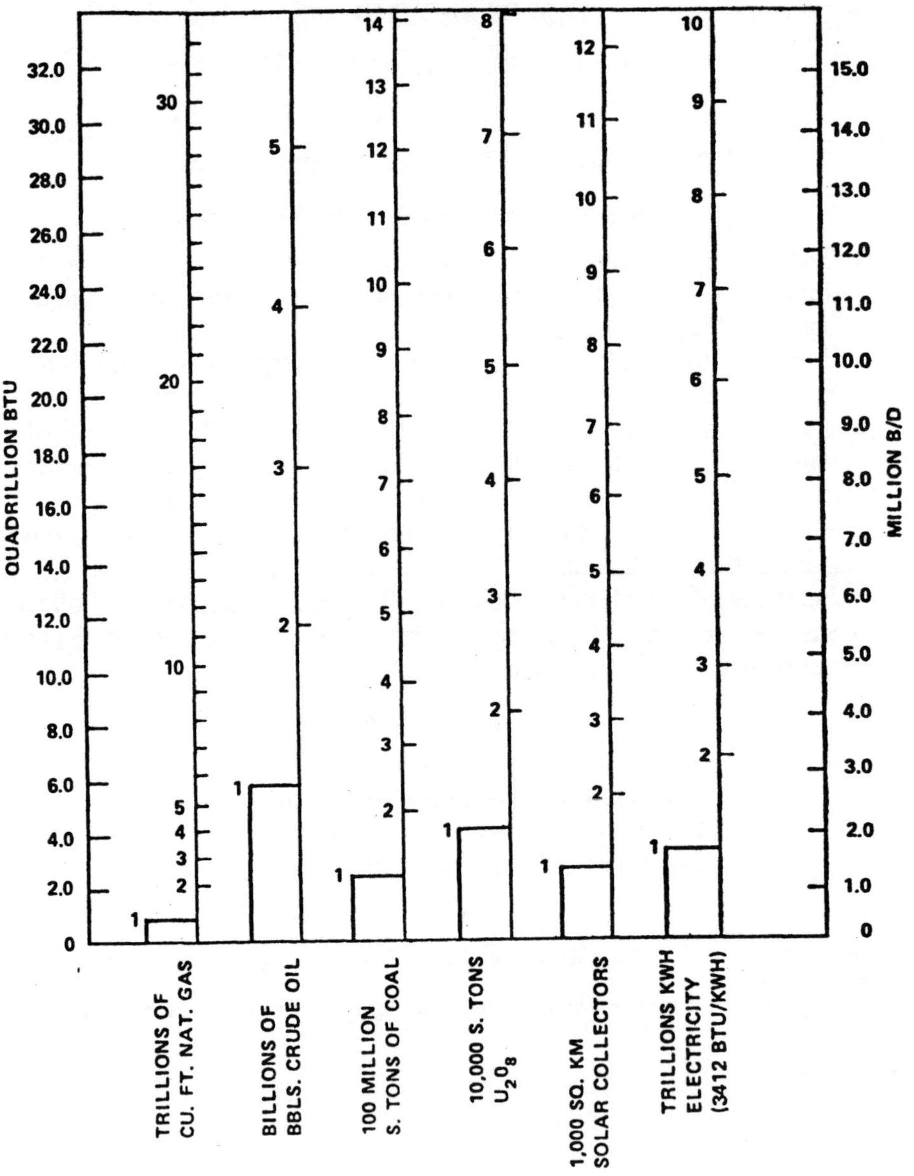

Table I.1.

ERDA SUMMARY FORECAST OF TOTAL U.S. ENERGY GROWTH (QUADRILLION Btu)

	1973	1980	1985	2000
Low (a)	75.6	86.1	96	135
Moderate (b)	75.6	89.7	105	174
High (c)	75.6	95.3	117	195

Growth Rate:

	1975-1985	1985-2000
Low	2.0%	2.3%
Moderate	2.75%	3.45%
High	3.7%*	3.5%

*cf. 3.6% 1965-1975 and 3.4% 1950-1975
(a) reflects high energy prices and conservation
(b) reflects some pricing and conservation effects.
(c) reflects return to low energy prices and no conservation

Source: R. W. A. LeGassie, ERDA, Testimony to the U.S. Committee on Interior and Insular Affairs, Subcommittee on Energy and the Environment, April 23, 1975.

Table I.2.

BREAKDOWN OF ERDA 1975 ELECTRICAL GENERATING CAPACITY PROJECTIONS

	Actual 1975	GWe installed at EOCY 1980	1985	1990
Low:				
Nuclear	38	70	160	285
IC/GT (a)	43	51	57	64
Fossil	347	411	482	532
Other (b)	64	72	86	99
Total	492	604	785	980
Moderate-Low:				
Nuclear	39	76	185	340
IC/GT	44	53	60	70
Fossil	348	416	465	523
Other	65	75	90	107
Total	496	620	800	1,040
Moderate-High:				
Nuclear	41	82	205	385
IC/GT	44	53	60	70
Fossil	350	420	465	513
Other	65	75	90	107
Total	500	630	820	1,075
High:				
Nuclear	43	92	245	470
IC/GT	44	55	64	75
Fossil	352	430	472	521
Other	66	77	94	114
Total	505	654	875	1,180

(a) internal combustion/gas turbine
(b) hydro/pumped storage

Source: R. W. A. LeGassie, ERDA, Testimony to the U.S. Committee on Interior and Insular Affairs, Subcommittee on Energy and the Environment, April 23, 1975.

Table I.3.

COMPARISON OF USAEC NUCLEAR POWER FORECASTS 1962-1975

AEC Forecast	Installed Nuclear Power at EOCY, GWe		
Made in Year	1975	1980	1985
1962 (a)	16	40	—
1964 (b)	29	75	—
1966 (c)	40	95	—
1967 (d)	61	145	255
1969 (e)	62	149	277
1970 (f)	59	150	300
1972 (g)	54	132	280
1973 (h)	47	102	255
1975 (i)	40	82	205

(a) Table 16 of Appendix IV of AEC Report to the President, "Civilian Nuclear Power", Dec. 1962.
(b) WASH-1055, March 1965.
(c) AEC Press Releases S-20-66 (June 7, 1966) & S-23-66 (Sept. 8, 1966).
(d) WASH-1084, December 1967.
(e) WASH-1139 statement of May 1969.
(f) WASH-1139, January 1971.
(g) WASH-1139 (72), December 1972.
(h) WASH-1139 (74), February 1974; mean of cases B&D.
(i) R. W. A. LeGassie, ERDA, Testimony to the U.S. Committee on Interior and Insular Affairs, Subcommittee on Energy and the Environment, April 23, 1975.

Source of comparison: Edison Electric Institute Nuclear Fuels Supply Study Program, 1976.

Table I.4.

SUMMARY PROJECTION OF UNITED STATES, FOREIGN AND WORLD INSTALLED NUCLEAR POWER GROWTH (GWe)

	1975	1980	1985	1990	2000
United States:					
High	—	85.3	204	389	1,005
Moderate	39.5	77.3	185	340	805
Low	—	72.9	157	213	507
Foreign:					
Non-Communist	33.9	114.8	303	571	1,451
Communist	7.5	26.3	81.8	209	729
World (Moderate)	80.9	218.4	569.8	1,120	2,985

Source: Edison Electric Institute Nuclear Fuels Supply Study Program, 1976.

Table I.5.

PROJECTED NUCLEAR POWER GROWTH FOR VARIOUS GEOPOLITICAL REGIONS (GWe)

Groups	Countries	1980	1985	1990	2000
	USA (a)	77	185	339	805
	Canada	7	18	41	115
EC (b)		59	155.3	284.8	632.6
	Belgium	3.5	9.5	16.5	30
	Denmark	—	1.8	4.9	11.4
	France	20.4	56*	90	170
	Germany	19.1	44.6	77	134
	Ireland	—	0.7	2	6
	Italy	6.0	23.0	54.7	140
	Luxembourg	—	1.2	1.2	1.2
	Netherlands	0.5	3.5	7.5	16
	UK	9.5	15.0	31	124
Pacific Asia		24.9	61	105.2	230
	Japan	16	50	90	190
Western Europe (c)		81	208	370	808
Communist Countries		26	82	209	729
World		218	570	1,119	2,985

(a) Moderate growth
(b) EC means European Community
(c) European OECD nations
*"Recent discussions with the French Government indicate this should now read 45-48"

Source: EEI (Edison Electric Institute Nuclear Fuels Supply Study Program, 1976)

Table I.6.

TOTAL PRIMARY ENERGY REQUIREMENTS AND AVERAGE ANNUAL ENERGY CONSUMPTION ESTIMATED GROWTH RATES FOR THE BASE PROJECTION*, $7.20 AND $10.80 CASES FOR THE PERIOD 1972-85**

	U.S.A.	Canada	E.E.C.	OECD Europe	Japan	Australia (1)	New Zealand (1)	Total OECD
TPER(2) 1972	1769.2	155.1	958.7	1157.5	318.5	55.8	7.4	3463.3
1980:								
Base Projection TPER	2357.8	234.8	1412.8	1729.5	636.4	97.7	11.4	5067.6
Average Growth Rate 1972-80 (%)	3.7	5.3	5.0	5.1	9.0	7.3	5.5	4.9
$7.20 case TPER	2247.4	220.4	1336.8	1632.4	576.7			4786.0
% Decrease from Base Projection	−4.7	−6.1	−5.4	−5.6	−9.4			−5.6
Average Growth Rate 1972-80 (%)	3.0	4.5	4.2	4.4	7.7			4.1
$10.80 Case TPER	2163.6	211.5	1278.4	1561.5	554.5			4600.2
% Decrease from Base Projection	−8.2	−9.9	−9.5	−9.7	−12.9			−9.2
Average Growth Rate 1972-80 (%)	2.5	4.0	3.7	3.8	7.2			3.6
1985								
Base Projection TPER	2881.0	292.9	1785.2	2244.9	854.1	133.4	14.2	6420.5
Average Growth Rate 1972-85 (%)	3.8	5.0	4.9	5.2	7.9	6.9	5.1	4.9
$7.20 Case TPER	2667.0	272.4	1686.4	2121.1	758.5			5966.6
% Decrease from Base Projection	−7.4	−7.0	−5.5	−5.5	−11.2			−7.1
Average Growth Rate 1972-85 (%)	3.2	4.4	4.4	4.8	6.9			4.3
$10.80 Case TPER	2522.4	248.8	1607.3	2022.7	708.5			5650.0
% Decrease from Base Projection	−12.4	−15.1	−10.0	−9.9	−17.0			−12.0
Average Growth Rate 1972-85 (%)	2.8	3.7	4.1	4.4	6.3			3.8

(1) No alternative cases.
(2) Total primary Energy Requirements in Mtoe (10^{13} Kcal).
* By OECD, prior to October, 1973, oil embargo and subsequent price increases.
**From preliminary OECD data, November, 1974.

Source: "Financing Free World Energy Supply and Use"; John E. Gray; Atlantic Council; 1975.

Table I.7

EXTRACT FROM
"POTENTIAL MARKET FOR NUCLEAR PLANTS (1981-1990): NUMBER OF PLANTS vs SIZE"

Country	Market[1] 1000MW	Reactor Size MW 200	300	400	600	800 and Larger
Spain	20.0					17
Mexico	19.6					19
Brazil	19.4				1	18
India	14.3			6		12
Yugoslavia	12.4				3	12
Korea	9.4				6	7
Iran	9.4				6	7
Czechoslovakia	7.8				5	6
Taiwan	7.4				7	4
Argentina	7.2				8	3
D.D.R.	7.2					8
Poland	7.0					7
Pakistan	5.9		1	5	6	
Egypt	5.6			6	4	1
Philippines (Luzon)	5.6				8	1
Turkey	5.6			2	8	
Bulgaria	5.4				6	
Venezuela	5.1		1		6	
Greece	5.1			1	6	
Romania	5.1		1		6	
Singapore	4.7		7	5	1	
Columbia	3.9			1	6	
Thailand	3.8		2	5	2	
Bangladesh	3.8		2	5	2	
Hungary	3.6				6	
Hong Kong	3.2			5	2	
Cuba	2.2		2	4		
Israel	2.1			5		
Peru	1.8		2	3		
Chile	1.6			4		
Jamaica	1.5	2	2			
Uruguay	1.0	5				
Malaysia (West)	1.7	1	5			
Indonesia (Java)	1.6	3	3			
Kuwait	1.2	6				
Iraq	0.8	4				
Nigeria	9.6	1				
Rep. Viet Nam	0.6	2				
TOTAL		24	28	57	105	122

[1] Totals do not reconcile in each case because reactors smaller than 200 MW are omitted in this Extract.

Source: IAEA data as presented in *Nuclear Power: Its Significance for the Developing World*, July 10, 1975; International Bank for Reconstruction and Development (World Bank). See also *Market Survey for Nuclear Power in Developing Countries*, IAEA, Vienna; 1974.

Table I.8.

SUMMARY OF THE NUCLEAR STATUS OF COUNTRIES HAVING AT LEAST ONE NUCLEAR REACTOR OR ONE ELEMENT OF THE NUCLEAR FUEL CYCLE IN THEIR TERRITORY

Country	Power reactors 1975	Power reactors 1980	Uranium enrichment capability (a)	Fuel reprocessing capability (a)	Uranium resources <$30/lb (b)	Uranium producer 1975/76	Research reactor in operation (c)	Breeder reactor programme
Angola					+			
Argentina	+	+			+	+	+	
Australia			P	P	+	+	+	
Austria		+					+	
Belgium	+	+		p (h)			+	
Brazil		+	P	P	+		+	
Bulgaria	+	+					+	
Canada	+	+	P		+	+	+	
Central African Republic					+			
Chile							+	
Columbia					+		+	
Czechoslovakia	+	+		O	+		+	
Denmark					+		+	
Egypt							+	
Finland		+			+		+	
France	+	+	O/C/Pp	O/C/Pp	+	+	+	+
Gabon					+	+		
Germany (DR)	+	+					+	
Germany (FR)	+	+	O/C	O/C/P	+		+	+
Greece							+	
Hungary		+					+	
India	+	+		O/C	+	+	+	+
Indonesia					+		+	
Iran		+					+	
Iraq							+	
Israel							+	
Italy	+	+		Pp	+	+	+	+
Japan	+	+	P	O/C/P	+	+	+	+
Korea, South		+		P			+	
Mauritania					+			
Mexico		+			+	+	+	
Netherlands	+	+	O/p				+	
Niger					+	+		
Norway							+	
Pakistan	+	+			+		+	
Philippines							+	
Poland							+	
Portugal					+		+	
Romania		+					+	
South Africa			O/P		+	+	+	
Spain	+	+		O	+	+	+	
Sweden	+	+	P		+	+	+	
Switzerland	+	+					+	
Taiwan		+					+	
Thailand							+	
Turkey					+		+	
UK	+	+	O/p	O/C/p			+	+
Uruguay							+	
USA	+	+	O/Pp	C/Pp	+	+	+	+
USSR	+	+	O	O	+	+	+	+
Venezuela			P				+	
Yugoslavia		+			+		+	
Zaire			P		+		+	

Table I.8 is continued on next page

Table 1.8. Continued

SUMMARY OF THE NUCLEAR STATUS OF COUNTRIES HAVING AT LEAST ONE NUCLEAR REACTOR OR ONE ELEMENT OF THE NUCLEAR FUEL CYCLE IN THEIR TERRITORY

Country	NPT status (d)	NPT Safeguards agreement (e)	Non-NPT safe-guards agreement with IAEA (c)	Member of IAEA (c)	Member of Euratom (f)	Member of NEA (g)
Angola						
Argentina			+	+		
Australia	R	*		+		+
Austria	R	*		+		+
Belgium	R	S		+	+	+
Brazil			+	+		
Bulgaria	R	*		+		
Canada	R	*		+		+
Central African Republic	R					
Chile			+	+		
Columbia	S		+	+		
Czechoslovakia	R	*		+		
Denmark	R	*		+	+	+
Egypt	S			+		
Finland	R	*		+		
France		nw		+	+	+
Gabon	R			+		
Germany (DR)	R	*		+		
Germany (FR)	R	S		+	+	+
Greece	R	*		+		+
Hungary	R	*		+		
India			+	+		
Indonesia	S	.	+	+		
Iran	R	*		+		
Iraq	R	*		+		
Israel			+	+		
Italy	R	S		+	+	+
Japan	S		+	+		+
Korea, South	R	*		+		
Mauritania						
Mexico	R	*		+		
Netherlands	R	S		+	+	+
Niger				+		
Norway	R	*		+		+
Pakistan			+	+		
Philippines	R	*		+		
Poland	R	*				
Portugal			+	+		+
Romania	R	*		+		
South Africa			+	+		
Spain			+	+		+
Sweden	R	*		+		+
Switzerland	S		+	+		+
Taiwan	R		+			
Thailand	R	*		+		
Turkey	S		+	+		+
UK	R	nw		+	+	+
Uruguay	R	S	+	+		
USA	R	nw		+		
USSR	R	nw		+		
Venezuela	R		+	+		
Yugoslavia	R	*		+		
Zaire	R	*		+		

Source: Stockholm International Peace Research Institute
Footnotes: please see opposite.

Footnotes:

(a) Commercial- or pilot-scale facility on country's territory: O = in operation; C = under construction; P = facility planned or under consideration; p = additional capacity planned for existing facility.
(b) The Nuclear Energy Agency (NEA) has defined world uranium resources in two ways: first according to the type of resource in geological terms, second according to a hypothetical market price. Geologically, ore deposits are classified as *reasonably assured resources* (RAR) or *estimated additional resources* (EAR), the difference being the reliability of the geological estimate. There are three price categories (per pound U_3O_8): (a) less than $15; (b) $15 to $30; and (c) $30 to $100 (these categories correspond to $10, $10-15 and $15-30 in the NEA's 1973 report). The recent escalation of the market price of uranium emphasizes that the NEA price levels should not be interpreted too literally in making economic assessments of nuclear power costs but they nevertheless serve as a useful indication of the competitive value of a country's uranium resources.
(c) As of 31 December 1975.
(d) As of 31 December 1975. R = ratified; S = signed.
(e) As of December 1975. * = in force; S = signed; nw = nuclear-weapon state.
(f) Euratom = European Atomic Energy Community.
(g) NEA = Nuclear Energy Agency of the Organization for Economic Cooperation and Development (OECD).
(h) The Eurochemic reprocessing plant in Mol has been shut down and future reopening is doubtful, but under consideration.

Sources: Facts on Nuclear Proliferation, a handbook prepared for the Committee on Government Operations, US Senate, by the Congressional Research Service, Library of Congress (Washington, US Printing Office, 1975) pp. 105-107 and 127-129; *Oversight Hearings on Nuclear Energy—International Proliferation of Nuclear Technology*, Hearings before the Subcommittee on Energy and the Environment of the Committee on Interior and Insular Affairs, US House of Representatives, 21, 22 and 24 July 1975 (Washington, US Government Printing Office, 1975) pp. 42-43; Poole, L. G., "World Uranium Resources", *Nuclear Engineering International*, Vol. 20, No. 224, February 1975, pp. 95-100; "The Nuclear Fuel Cycle", *Nuclear Engineering International*, Vol. 20, No. 237, December, 1975, pp. 1015-1020; Rippon, Simon, "Reprocessing—What Went Wrong?, *Nuclear Engineering International*, Vol. 21, No. 239, February 1976, pp. 21-27.

Table I.9.

NUCLEAR STATUS OF COUNTRIES HAVING AT LEAST ONE NUCLEAR REACTOR OR ONE ELEMENT OF THE NUCLEAR FUEL CYCLE IN THEIR TERRITORY

ARGENTINA
Year of operation of 1st power reactor: 1974.
Number of power reactors 1974: 1.
Number of power reactors 1980: 2.
Dominant reactor type: HWR.
Total output of power reactors (net MWe) 1974: 320.
Total output of power reactors (net MWe) 1980: 920.
Approximate annual production of plutonium (kg) 1980: 270.
Approximate accumulated stock of plutonium (kg) 1980: 1000.
Year of operation of 1st research reactor: 1958.
Number of research ractors in operation 1974: 5.
Uranium resources (tons): <$10 lb - 23,000; >$10/lb - 31,000.
Planned uranium production (tons) 1975: 165.
Member of IAEA.
NPT status: not a member.
Non-NPT safeguards agreement signed with IAEA.
Military expenditure 1973: $887 mn.
Nuclear-capable delivery systems: Aircraft—63 Douglas A-4 Skyhawk; 11 Canberra; 12 Mirage III E.

AUSTRALIA
Year of operation of 1st research reactor: 1958.
Number of research reactors in operation 1974: 2.
Uranium resources (tons): <$10/lb - 149,000; >$10/lb - 58,000.
Planned uranium production (tons) 1975: 770.
Member of IAEA.
NPT status: ratified.
NPT safeguards agreement in force.
Nuclear-capable delivery systems: Aircraft—24 F-111 C; 8 Canberra; 15 Douglas A-4 Skyhawk; 100 Mirage III.

AUSTRIA
Year of operation of 1st power reactor: 1976.
Number of power reactors 1980: 1.
Dominant reactor type: LWR.
Total output of power reactors (net MWe) 1980: 690.
Approximate annual production of plutonium (kg) 1980: 170.
Approximate accumulated stock of plutonium (kg) 1980: 850.
Year of operation of 1st research reactor: 1960.
Number of research reactors in operation 1974: 3.
Member of IAEA.
NPT status: ratified.
NPT safeguards agreement in force.

BELGIUM
Year of operation of 1st power reactor: 1974.
Number of power reactors 1974: 1.
Number of power reactors 1980: 3.
Dominant reactor type: LWR.
Total output of power reactors (net MWe) 1974: 390.
Total output of power reactors (net MWe) 1980: 1650.

Approximate annual production of plutonium (kg) 1980: 400.
Approximate accumulated stock of plutonium (kg) 1980: 2500.
Year of operation of 1st research reactor: 1956.
Number of research reactors in operation 1974: 5.
Reprocessing facilities: Eurochemic plant since 1966 - capacity 100,000 kg/year (joint with NEA).
Member of IAEA.
Member of Euratom.
NPT status: signed, not ratified.
NPT safeguards agreement signed.
Military expenditure 1973: $1263 mn.
Nuclear-capable delivery systems: Aircraft—60 F-104 G Starfighter; 80 Mirage V.

BRAZIL

Year of operation of 1st power reactor: 1976.
Number of power reactors 1980: 1.
Dominant reactor type: LWR.
Total output of power reactors (net MWe) 1980: 600.
Approximate annual production of plutonium (kg) 1980: 150.
Approximate accumulated stock of plutonium (kg) 1980: 750.
Year of operation of 1st research reactor: 1957.
Number of research reactors in operation 1974: 3.
Uranium resources (tons): <$10/lb - 2500; >$10/lb - 700.
Member of IAEA.
NPT status: not a member.
Non-NPT safeguards agreement signed with IAEA.
Nuclear-capable delivery systems: Aircraft—12 Mirage III E.

BULGARIA

Year of operation of 1st power reactor: 1974.
Number of power reactors 1974: 1.
Number of power reactors 1980: 4.
Dominant reactor type: LWR.
Total output of power reactors (net MWe) 1974: 440.
Total output of power reactors (net MWe) 1980: 1760.
Approximate annual production of plutonium (kg) 1980: 440.
Approximate accumulated stock of plutonium (kg) 1980: 1650.
Year of operation of 1st research reactor: 1961.
Number of research reactors in operation 1974: 1.
Member of IAEA.
NPT status: ratified.
NPT safeguards agreement in force.
Military expenditure 1973: $364 mn.
Nuclear-capable delivery systems: Aircraft—12 Ilyushin-28; Missiles—18 Scud SSM.

CANADA

Years of operation of 1st power reactor: 1962.
Number of power reactors 1974: 7.
Number of power reactors 1980: 12.
Dominant reactor type: HWR.
Total output of power reactors (net MWe) 1974: 2510.
Total output of power reactors (net MWe) 1980: 6120.
Approximate annual production of plutonium (kg) 1974: 750.
Approximate annual production of plutonium (kg) 1980: 2000.
Approximate accumulated stock of plutonium (kg) 1974: 3000.

Approximate accumulated stock of plutonium (kg) 1980: 12,000.
Year of operation of 1st research reactor: 1947.
Number of research reactors in operation 1974: 8.
Uranium resources (tons): <$10/lb - 375,000; >$10/lb - 341,000.
Planned uranium production (tons) 1975: 6500.
Member of IAEA.
NPT status: ratified.
NPT safeguards agreement in force.
Military expenditure 1973: $2391 mn.
Nuclear-capable delivery systems: Aircraft—50 CF-104 D Starfighter.

CENTRAL AFRICAN REPUBLIC
Uranium resources (tons): <$10/lb - 16,000.
NPT status: ratified.

CHILE
Year of operation of 1st research reactor: 1973.
Number of research reactors in operation 1974: 1.
Member of IAEA.
NPT status: not a member.
Non-NPT safeguards agreement signed with IAEA.

CHINA
Has no peaceful nuclear power programme, so far as is known. It has uranium resources. Its enriched-uranium plant, reprocessing plant, and nuclear reactor are part of its nuclear-weapon programme. This programme has, of course, given China considerable experience in nuclear technology.
NPT status: not a member.
Nuclear-weapon state.

COLOMBIA
Year of operation of 1st research reactor: 1965.
Number of research reactors in operation 1974: 1.
Member of IAEA.
NPT status: signed.
Non-NPT safeguards agreement signed with IAEA.

CZECHOSLOVAKIA
Year of operation of 1st power reactor: 1972.
Number of power reactors 1974: 1.
Number of power reactors 1980: 5.
Dominant reactor type: LWR.
Total output of power reactors (net MWe) 1974: 110.
Total output of power reactors (net MWe) 1980: 1760.
Approximate annual production of plutonium (kg) 1974: 25.
Approximate annual production of plutonium (kg) 1980: 440.
Approximate accumulated stock of plutonium (kg) 1974: 75.
Approximate accumulated stock of plutonium (kg) 1980: 1200.
Year of operation of 1st research reactor: 1957.
Number of research reactors in operation 1974: 3.
Uranium resources: yes - but amount not known.
Member of IAEA.
NPT status: ratified.
NPT safeguards agreement in force.
Military expenditure 1973: $1965 mn.
Nuclear-capable delivery systems: Aircraft—20 Ilyushin-28; 56 Su-7 "Fitter"; Missiles—27 Scud SSM.

DENMARK
Year of operation of 1st research reactor: 1957.
Number of research reactors in operation 1974: 3.
Uranium resources (tons): (Greenland) <$10/lb - 15,600.
Member of IAEA.
Member of Euratom.
NPT status: ratified.
NPT safeguards agreement in force.
Military expenditure 1973: $615 mn.
Nuclear-capable delivery systems: Aircraft—40 F-104 G Starfighter; 40 F-100 D/F Super Sabre.

EGYPT
Year of operation of 1st research reactor: 1961.
Number of research reactors in operation 1974: 1.
Member of IAEA.
NPT status: signed.
Military expenditure 1973: $2818 mn.
Nuclear-capable delivery systems: Aircraft—35-40 MiG-21 MF "Fishbed"; 25 Tu-16 "Badger"; 5 Ilyushin-28; 100 Su-7 "Fitter"; Missiles—24 Scud SSM.

FINLAND
Year of operation of 1st power reactor: 1976.
Number of power reactors 1980: 3.
Dominant reactor type: LWR.
Total output of power reactors (net MWe) 1980: 1540.
Approximate annual production of plutonium (kg) 1980: 400.
Approximate accumulated stock of plutonium (kg) 1980: 1400.
Year of operation of 1st research reactor: 1962.
Number of research reactors in operation 1974: 1.
Uranium resources (tons): <$10/lb - 1300.
Member of IAEA.
NPT status: ratified.
NPT safeguards agreement in force.
Military expenditure 1973: $230 mn.

FRANCE
Year of operation of 1st power reactor: 1958.
Number of power reactors 1980: 23 (1).
Dominant reactor type: Graphite/LWR.
Total output of power reactors (net MWe) 1974: 2870.
Total output of power reactors (net MWe) 1980: 15,170.
Approximate annual production of plutonium (kg) 1974: 700.
Approximate annual production of plutonium (kg) 1980: 4000.
Approximate accumulated stock of plutonium (kg) 1974: 4500.
Approximate accumulated stock of plutonium (kg) 1980: 18,000.
Year of operation of 1st research reactor: 1948.
Number of research reactors in operation 1974: 23.
Breeder reactor developments: research breeder in operation (1967); 250 MWe breeder in operation (1973); 1200 MWe breeder planned (1979).
Uranium resources (tons): <$10/lb - 61,000; >$10/lb - 45,000.
Planned uranium production (tons) 1975: 1800.
Reprocessing facilities: Marcoule plant since 1958 - capacity about 1,000,000 kg/yr; La Hague plant since 1958 - capacity about 1,000,000 kg/yr.

Enrichment plants and plans: gas diffusion plant at Pierrelatte - capacity 300 tons SW/yr.
Member of IAEA.
Euratom member.
NPT status: not a member.
Nuclear-weapon state.

GABON
Uranium resources (tons): <$10/lb - 25,000; >$10/lb - 5000.
Planned uranium production (tons) 1975: 600.
Member of IAEA.
NPT status: ratified.
NPT safeguards agreement under negotiation.

GERMANY FR
Year of operation of 1st power reactor: 1965.
Number of power reactors 1974: 10.
Number of power reactors 1980: 28 (4).
Dominant reactor type: LWR.
Total output of power reactors (net MWe) 1974: 4000.
Total output of power reactors (net MWe) 1980: 21,600.
Approximate annual production of plutonium (kg) 1974: 1000.
Approximate annual production of plutonium (kg) 1980: 5000.
Approximate accumulated stock of plutonium (kg) 1974: 3500.
Approximate accumulated stock of plutonium (kg) 1980: 24,000.
Year of operation of 1st research reactor: 1957.
Number of research reactors in operation 1974: 33.
Breeder reactor developments: 20 MWe breeder planned; 280 MWe breeder under construction.
Reprocessing facilities: Karlsruhe plant since 1971 - capacity 40,000 kg/yr.
Enrichment plants and plans: cooperating in UK/Netherlands/FRG tripartite centrifuge plants; developing jet nozzle method at Karlsruhe.
Member of IAEA.
Member of Euratom.
NPT status: signed, not ratified.
NPT safeguards agreement signed.
Military expenditure 1973: $11,910 mn.
Nuclear-capable delivery systems: Aircraft—308 F-104 G Starfighter; 60 F-4 F Phantom II; Missiles—Lance SSM; 72 Pershing 1A SSM; 19 Sargent SSM.

GERMANY DR
Year of operation of 1st power reactor: 1966.
Number of power reactors 1974: 2.
Number of power reactors 1980: 3.
Dominant reactor type: LWR.
Total output of power reactors (net MWe) 1974: 430.
Total output of power reactors (net MWe) 1980: 800.
Approximate annual production of plutonium (kg) 1974: 100.
Approximate annual production of plutonium (kg) 1980: 200.
Approximate accumulated stock of plutonium (kg) 1974: 600.
Approximate accumulated stock of plutonium (kg) 1980: 1100.
Year of operation of 1st research reactor: 1957.
Number of research reactors in operation 1974: 1.
Member of IAEA.
NPT status: ratified.
NPT safeguards agreement in force.

Military expenditure 1973: $2457 mn.
Nuclear-capable delivery systems: Missiles—Scud SSM.

GREECE
Year of operation of 1st research reactor: 1961.
Number of research reactors in operation 1974: 1.
Member of IAEA.
NPT status: ratified.
NPT safeguards agreement provisionally in force.

HUNGARY
Year of operation of 1st power reactor: 1980.
Number of power reactors 1980: 1 (1).
Dominant reactor type: LWR.
Total output of power reactors (net MWe) 1980: 440.
Approximate annual production of plutonium (kg) 1980: 100.
Approximate accumulated stock of plutonium (kg) 1980: 100.
Year of operation of 1st research reactor: 1959.
Number of research reactors in operation 1974: 2.
Member of IAEA.
NPT status: ratified.
NPT safeguards agreement in force.
Military expenditure 1973: $567 mn.
Nuclear-capable delivery systems: Aircraft—12 Su-7 "Fitter"; Missiles—12 Scud SSM.

INDIA
Year of operation of 1st power reactor: 1969.
Number of power reactors 1974: 4.
Number of power reactors 1980: 8 (2).
Dominant reactor type: LWR.
Total output of power reactors (net MWe) 1974: 780.
Total output of power reactors (net MWe) 1980: 1580.
Approximate annual production of plutonium (kg) 1974: 200.
Approximate annual production of plutonium (kg) 1980: 400.
Approximate accumulated stock of plutonium (kg) 1974: 700.
Approximate accumulated stock of plutonium (kg) 1980: 2500.
Year of operation of 1st research reactor: 1956.
Number of research reactors in operation 1974: 4.
Breeder reactor developments: 30 MWe experimental breeder under construction.
Uranium resources (tons): <$10/lb - 3100. Substantial thorium.
Reprocessing facilities: Tarapur plant; also facility at the Bhabha Atomic Research Centre.
Member of IAEA.
NPT status: not a member.
Non-NPT safeguards agreement signed with IAEA.
Military expenditure 1973: $2402 mn.
Nuclear-capable delivery systems: Aircraft—60 Canberra; 77 Su-7 "Fitter"; some MiG-21 MF "Fishbed".

INDONESIA
Year of operation of 1st research reactor: 1964.
Number of research reactors in operation 1974: 1.
Member of IAEA.
NPT status: signed.
Non-NPT safeguards agreement signed with IAEA.

IRAN
Year of operation of 1st research reactor: 1967.
Number of research reactors in operation 1974: 1.
Member of IAEA.
NPT status: ratified.
NPT safeguards agreement in force.

IRAQ
Year of operation of 1st research reactor: 1968.
Number of research reactors in operation 1974: 1.
Member of IAEA.
NPT status: ratified.
NPT safeguards agreement in force.

ISRAEL
Year of operation of 1st research reactor: 1960.
Number of research reactors in operation 1974: 2.
Uranium resources: some related to phosphate production.
Member of IAEA.
NPT status: not a member.
Non-NPT safeguards agreement signed with IAEA.
Military expenditure 1973: $3051 mn.
Nuclear-capable delivery systems: Aircraft—150 F-4 E Phantom II; 180 A-4 E/H Skyhawk; 10 Vautour; Missiles—Jericho SSM.

ITALY
Year of operation of 1st power reactor: 1962.
Number of power reactors 1974: 3.
Number of power reactors 1980: 7 (1).
Dominant reactor type: LWR.
Total output of power reactors (net MWe) 1974: 610.
Total output of power reactors (net MWe) 1980: 3380.
Approximate annual production of plutonium (kg) 1974: 200.
Approximate annual production of plutonium (kg) 1980: 900.
Approximate accumulated stock of plutonium (kg) 1974: 2000.
Approximate accumulated stock of plutonium (kg) 1980: 5000.
Year of operation of 1st research reactor: 1959.
Number of research reactors in operation 1974: 16.
Breeder reactor developments: research breeder planned.
Uranium resources (tons): <$10/lb - 1200; >$10/lb - 4200.
Planned uranium production (tons) 1975: 92.
Reprocessing facilities: Eurex-1 plant since 1970 - capacity 25,000 kg/yr.
Member of IAEA.
Member of Euratom.
NPT status: signed, not ratified.
NPT safeguards agreement signed.
Military expenditure 1973: $4094 mn.
Nuclear-capable delivery systems: Aircraft—194 F-104 G/S Starfighter.

JAPAN
Year of operation of 1st power reactor: 1965.
Number of power reactors 1974: 10.
Number of power reactors 1980: 29 (3).
Dominant reactor type: LWR.
Total output of power reactors (net MWe) 1974: 5000.

Total output of power reactors (net MWe) 1980: 19,400.
Approximate annual production of plutonium (kg) 1974: 1200.
Approximate annual production of plutonium (kg) 1980: 5000.
Approximate accumulated stock of plutonium (kg) 1974: 3500.
Approximate accumulated stock of plutonium (kg) 1980: 25,000.
Year of operation of 1st research reactor: 1960.
Number of research reactors in operation 1974: 21.
Breeder reactor developments: research breeder under construction; 300 MWe breeder planned.
Uranium resources (tons): <$10/lb - 2800; >$10/lb - 4200.
Planned uranium production (tons) 1975: 30.
Reprocessing facilities: Tokai plant since 1974 - capacity 260,000 kg/yr.
Member of IAEA.
NPT status: signed, not ratified.
Non-NPT safeguards agreement signed with IAEA.
Military expenditure 1973: $3366 mn.
Nuclear-capable delivery systems: Aircraft—40 F-4 E5 Phantom II; 130 F-104 J Starfighter.

KOREA, South
Year of operation of 1st power reactor: 1975.
Number of power reactors 1980: 2 (1).
Dominant reactor type: HWR.
Total output of power reactors (net MWe) 1980: 1160.
Approximate annual production of plutonium (kg) 1980: 300.
Approximate accumulated stock of plutonium (kg) 1980: 1000.
Year of operation of 1st research reactor: 1962.
Number of research reactors in operation 1974: 1.
Member of IAEA.
NPT status: signed, not ratified.
Non-NPT safeguards agreement signed with IAEA.
Military expenditure 1973: $453 mn.
Nuclear-capable delivery systems: Aircraft—30 F-4 D Phantom II.

MEXICO
Year of operation of 1st power reactor: 1977.
Number of power reactors 1980: 2.
Dominant reactor type: LWR.
Total output of power reactors (net MWe) 1980: 1300.
Approximate annual production of plutonium (kg) 1980: 350.
Approximate accumulated stock of plutonium (kg) 1980: 1100.
Year of operation of 1st research reactor: 1968.
Number of research reactors in operation 1974: 2.
Uranium resources (tons): <$10/lb - 1000; >$10/lb - 900.
Planned uranium production (tons) 1975: 225.
Member of IAEA.
NPT status: ratified.
NPT safeguards agreement in force.
Military expenditure 1973: $353 mn.

NETHERLANDS
Year of operation of 1st power reactor: 1968.
Number of power reactors 1974: 2.
Number of power reactors 1980: 2.
Dominant reactor type: LWR.
Total output of power reactors (net MWe) 1974: 530.

Total output of power reactors (net MWe) 1980: 530.
Approximate annual production of plutonium (kg) 1974: 130.
Approximate annual production of plutonium (kg) 1980: 130.
Approximate accumulated stock of plutonium (kg) 1974: 300.
Approximate accumulated stock of plutonium (kg) 1980: 1000.
Year of operation of 1st research reactor: 1959.
Number of research reactors in operation 1974: 6.
Enrichment plants and plans: 2 centrifuge pilot plants at Almelo - total capacity 50 tons SW/yr.
Member of IAEA.
Member of Euratom.
NPT status: signed, not ratified.
NPT safeguards agreement signed.
Military expenditure 1973: $2034 mn.
Nuclear-capable delivery systems: Aircraft—90 F-104 G Starfighter; Missiles—Lance SSM (on order).

NIGER
Uranium resources (tons): <$10/lb - 60,000; >$10/lb - 20,000.
Planned uranium production (tons) 1975: 1500.
Member of IAEA.
NPT status: not a member.

NORWAY
Year of operation of 1st research reactor: 1959.
Number of research reactors in operation 1974: 2.
Member of IAEA.
NPT status: ratified.
NPT safeguards agreement in force.

PAKISTAN
Year of operation of 1st power reactor: 1971.
Number of power reactors 1974: 1.
Number of power reactors 1980: 1.
Dominant reactor type: HWR.
Total output of power reactors (net MWe) 1974: 125.
Total output of power reactors (net MWe) 1980: 125.
Approximate annual production of plutonium (kg) 1974: 30.
Approximate annual production of plutonium (kg) 1980: 30.
Approximate accumulated stock of plutonium (kg) 1974: 120.
Approximate accumulated stock of plutonium (kg) 1980: 300.
Year of operation of 1st research reactor: 1965.
Number of research reactors in operation 1974: 1.
Member of IAEA.
NPT status: not a member.
Non-NPT safeguards agreement signed with IAEA.
Military expenditure 1973: $459 mn.
Nuclear-capable delivery systems: Aircraft—10 Canberra, 28 Mirage V, 21 Mirage III E.

PHILIPPINES
Year of operation of 1st research reactor: 1963.
Number of research reactors in operation 1974: 1.
Member of IAEA.
NPT status: ratified.
NPT safeguards agreement in force.

POLAND
Year of operation of 1st research reactor: 1958.
Number of research reactors in operation 1974: 4.
Member of IAEA.
NPT status: ratified.
NPT safeguards agreement in force.

PORTUGAL
Year of operation of 1st research reactor: 1961.
Number of research reactors in operation 1974: 1.
Member of IAEA.
NPT status: not a member.
Non-NPT safeguards agreement signed with IAEA.

ROMANIA
Year of operation of 1st research reactor: 1957.
Number of research reactors in operation 1974: 1.
Member of IAEA.
NPT status: ratified.
NPT safeguards agreement in force.

SOUTH AFRICA
Year of operation of 1st research reactor: 1965.
Number of research reactors in operation 1974: 2.
Uranium resources (tons): <\$10/lb - 210,000; >\$10/lb - 88,000.
Planned uranium production (tons) 1975: 3800.
Enrichment plants and plans: Plant at Pelindaba - probably based on jet nozzle method. Output not announced.
Member of IAEA.
NPT status: not a member.
Non-NPT safeguards agreement signed with IAEA.
Military expenditure 1973: \$650 mn.
Nuclear-capable delivery systems: Aircraft—6 Canberra, 10 Buccaneer, 48 Mirage F-1, 32 Mirage III E.

SPAIN
Year of operation of 1st power reactor: 1968.
Number of power reactors 1974: 3.
Number of power reactors 1980: 11 (2).
Dominant reactor type: LWR.
Total output of power reactors (net MWe) 1974: 1070.
Total output of power reactors (net MWe) 1980: 8550.
Approximate annual production of plutonium (kg) 1974: 250.
Approximate annual production of plutonium (kg) 1980: 2200.
Approximate accumulated stock of plutonium (kg) 1974: 1000.
Approximate accumulated stock of plutonium (kg) 1980: 8000.
Year of operation of 1st research reactor: 1958.
Number of research reactors in operation 1974: 5.
Uranium resources (tons): <\$10/lb - 8500; >\$10/lb - 7700.
Planned uranium production (tons) 1975: 132.
Member of IAEA.
NPT status: not a member.
Non-NPT safeguards agreement signed with IAEA.
Military expenditure 1973: \$1131 mn.
Nuclear-capable delivery systems: Aircraft—36 F-4C Phantom II, 24 Mirage III E.

SWEDEN
Year of operation of 1st power reactor: 1970.
Number of power reactors 1974: 4.
Number of power reactors 1980: 11 (1).
Dominant reactor type: LWR.
Total output of power reactors (net MWe) 1974: 2600.
Total output of power reactors (net MWe) 1980: 8300.
Approximate annual production of plutonium (kg) 1974: 650.
Approximate annual production of plutonium (kg) 1980: 2000.
Approximate accumulated stock of plutonium (kg) 1974: 1100.
Approximate accumulated stock of plutonium (kg) 1980: 8000.
Year of operation of 1st research reactor: 1960.
Number of research reactors in operation 1974: 3.
Uranium resources (tons): <$10/lb - 310,000.
Planned uranium production (tons) 1975: 120.
Member of IAEA.
NPT status: ratified.
NPT safeguards agreement under negotiation.
Non-NPT safeguards agreement signed with IAEA.
Military expenditure 1973: $1696 mn.

SWITZERLAND
Year of operation of 1st power reactor: 1969.
Number of power reactors 1974: 3.
Number of power reactors 1980: 8 (4).
Dominant reactor type: LWR.
Total output of power reactors (net MWe) 1974: 1000.
Total output of power reactors (net MWe) 1980: 5700.
Approximate annual production of plutonium (kg) 1974: 250.
Approximate annual production of plutonium (kg) 1980: 1400.
Approximate accumulated stock of plutonium (kg) 1974: 1200.
Approximate accumulated stock of plutonium (kg) 1980: 5000.
Year of operation of 1st research reactor: 1957.
Number of research reactors in operation 1974: 6.
Member of IAEA.
NPT status: signed, not ratified.
NPT safeguards agreement under negotiation.
Non-NPT safeguards agreement signed with IAEA.
Military expenditure 1973: $793 mn.

TAIWAN
Year of operation of 1st power reactor: 1976.
Number of power reactors 1980: 4.
Dominant reactor type: LWR.
Total output of power reactors (net MWe) 1980: 3100.
Approximate annual production of plutonium (kg) 1980: 700.
Approximate accumulated stock of plutonium (kg) 1980: 1400.
Year of operation of 1st research reactor: 1963.
Number of research reactors in operation 1974: 2.
NPT status: ratified.
Non-NPT safeguards agreement signed with IAEA.
Military expenditure 1973: $678 mn.
Nuclear-capable delivery systems: Aircraft—90 F-100 A/D Super Sabre; 63 F-104G Starfighter.

THAILAND
Year of operation of 1st power reactor: 1980.
Number of power reactors 1980: 1 (1).
Total output of power reactors (net MWe) 1980: 500.
Approximate annual production of plutonium (kg) 1980: 120.
Year of operation of 1st research reactor: 1962.
Number of research reactors in operation 1974: 1.
Member of IAEA.
NPT status: not a member.
NPT safeguards agreement in force.
Military expenditure 1973: $304 mn.
Nuclear-capable delivery systems: Aircraft—A-4B Skyhawk (on order).

TURKEY
Year of operation of 1st research reactor: 1962.
Number of research reactors in operation 1974: 1.
Uranium resources (tons): <$10/lb - 2200; >$10/lb - 500.
Member of IAEA.
NPT status: signed, not ratified.
Non-NPT safeguards agreement signed with IAEA.

USSR
Year of operation of 1st power reactor: 1958.
Number of power reactors 1974: 16.
Number of power reactors 1980: 24.
Dominant reactor type: Graphite.
Total output of power reactors (net MWe) 1974: 3370.
Total output of power reactors (net MWe) 1980: 10,000.
Approximate annual production of plutonium (kg) 1974: 850.
Approximate annual production of plutonium (kg) 1980: 2500.
Approximate accumulated stock of plutonium (kg) 1974: 5500.
Approximate accumulated stock of plutonium (kg) 1980: 20,000.
Year of operation of 1st research reactor: 1949.
Number of research reactors in operation 1974: 26.
Breeder reactor developments: 5 MWe research breeder in operation (1959); 12 MWe experimental breeder in operation (1965); 350 MWe breeder in operation (1972); 600 MWe breeder under construction.
Uranium resources: yes - amount not known.
Reprocessing facilities: yes - capacities not known.
Enrichment plants and plans: yes - capacities not known.
Member of IAEA.
NPT status: ratified.
Nuclear-weapon state.

UK
Year of operation of 1st power reactor: 1956.
Number of power reactors 1974: 31.
Number of power reactors 1980: 39.
Dominant reactor type: Graphite.
Total output of power reactors (net MWe) 1974: 5790.
Total output of power reactors (net MWe) 1980: 10,740.
Approximate annual production of plutonium (kg) 1974: 2000.
Approximate annual production of plutonium (kg) 1980: 3000.
Approximate accumulated stock of plutonium (kg) 1974: 15,000.
Approximate accumulated stock of plutonium (kg) 1980: 29,000.
Year of operation of 1st research reactor: 1947.

Number of research reactors in operation 1974: 24.
Breeder reactor developments: 15 MWe experimental breeder in operation (1959); 250 MWe breeder under construction; 1300 MWe breeder planned.
Reprocessing facilities: Windscale plant since 1964 - capacity 2,500,000 kg/yr natural U, 300,000 kg/yr enriched U; Douneray plant since 1958 - capacity 10,000 kg/yr highly enriched fuels.
Enrichment plants and plans: gas diffusion plant at Capenhurst - capacity 400 tons SW/yr. Pilot gas centrifuge plant at Capenhurst - capacity 25 tons SW/yr.
Member of IAEA.
Member of Euratom.
NPT status: ratified.
Nuclear-weapon state.

USA
Years of operation of 1st power reactor: 1957.
Number of power reactors 1974: 60.
Number of power reactors 1980: 156 (37).
Dominant reactor type: LWR.
Total output of power reactors (net MWe) 1974: 40,400.
Total output of power reactors (net MWe) 1980: 137,800.
Approximate annual production of plutonium (kg) 1974: 10,000.
Approximate annual production of plutonium (kg) 1980: 35,000.
Approximate accumulated stock of plutonium (kg) 1974: 30,000.
Approximate accumulated stock of plutonium (kg) 1980: 160,000.
Year of operation of 1st research reactor: 1950.
Number of research reactors in operation 1974: 117.
Breeder reactor developments: Research breeder in operation (1969); research breeder under construction; 18.5 MWe experimental breeder in operation (1965); 60 MWe breeder in operation (1965); 2400 MWe breeder planned.
Uranium resources (tons): <10/lb - 797,000, plus 70,000 tons uranium by-product from phosphate and copper production; >\$10/lb - 372,000.
Planned uranium production (tons) 1975: 14,600.
Reprocessing facilities: West Valley plant, New York, since 1966 - capacity about 200,000 kg/yr; Morris plant, Illinois, since 1971 - capacity about 300,000 kg/yr.
Enrichment plants and plans: three gas diffusion plants - total maximum capacity 17,000 tons SW/yr.
Member of IAEA.
NPT status: ratified.
Nuclear-weapon state.

URUGUAY
Year of operation of 1st research reactor: 1973.
Number of research reactors in operation 1974: 1.
Member of IAEA.
NPT status: ratified.
NPT safeguards agreement signed.
Non-NPT safeguards agreement signed with IAEA.

VENEZUELA
Year of operation of 1st research reactor: 1960.
Number of research reactors in operation 1974: 1.
Member of IAEA.
NPT status: signed.
NPT safeguards agreement signed with IAEA.

VIETNAM
Year of operation of 1st research reactor: 1963.
Number of research reactors in operation 1974: 1.
Member of IAEA.
NPT status: ratified.
NPT safeguards agreement in force.

YUGOSLAVIA
Year of operation of 1st power reactor: 1977.
Number of power reactors 1980: 2 (1).
Dominant reactor type: LWR.
Total output of power reactors (net MWe) 1980: 1400.
Approximate annual production of plutonium (kg) 1980: 350.
Approximate accumulated stock of plutonium (kg) 1980: 800.
Year of operation of 1st research reactor: 1958.
Number of research reactors in operation 1974: 3.
Uranium resources (tons): <$10/lb - 16,000.
Member of IAEA.
NPT status: ratified.
NPT safeguards agreement in force.
Military expenditure 1973: $792 mn.

ZAIRE
Year of operation of 1st research reactor: 1950.
Number of research reactors in operation 1974: 1.
Uranium resources (tons): <$10/lb - 3500.
Member of IAEA.
NPT status: ratified.
NPT safeguards agreement in force.

Notes:

Power reactors—output greater than 20 MWe/sec. The number in brackets after the 1980 figure is that of planned reactors not under construction on 1 January 1975.
Uranium resources—reasonably assured resources plus additional resources plus estimated additional resources.
NPT status—as of 1 January 1975.
Non-NPT safeguards agreement—as of 1 January 1975.
Military expenditure and nuclear-capable delivery systems—the UN Secretary General's expert study of nuclear weapons (1967) estimated the total 10-year costs of a small unsophisticated nuclear force to be $1700 million and the 10-year costs of a high-quality force to be $5600 million. The British and French nuclear forces have so far each involved expenditures of about $10,000 million. It is reported that the Indian nuclear explosion cost $400,000, including the cost of the plutonium and the test site. Many non-nuclear-weapon countries could afford to finance a nuclear force from current military expenditure without diverting a large fraction of this expenditure from other uses. Countries with a nuclear-capable delivery system could, of course acquire a nuclear force for considerably less cost.
SSM—Surface-to-surface missile.

Source: SIPRI, op. cit., pages 63 to 75.

Table II.1.

UNITED STATES TOTAL GROSS CONSUMPTION OF ENERGY RESOURCES BY MAJOR SOURCES AND CONSUMING SECTORS, 1971 ACTUAL, AND PROJECTIONS TO YEAR 2000

In Trillions of BTUs

	Coal[1]	Petroleum[2]	Natural gas	Total fossil fuels	Nuclear power	Hydro power	Total gross energy inputs	Synthetic gas distributed	Total four sector inputs	Utility elec. distributed	Total three sector inputs
1971 Household & Commercial	390	6,545	7,346	14,281	—	—	14,281	—	14,281	3,160	17,441
Industrial	4,465	5,391	10,438	20,294	—	—	20,294	—	20,294	2,329	22,623
Transportation	7	16,139	825	16,971	—	—	16,971	—	16,971	18	16,989
Electrical Generation	7,698	2,417	4,125	14,240	405	2,798	17,443	—	17,443	(5,507)	—
Synthetic Gas	—	—	—	—	—	—	—	—	—	—	—
Total	12,560	30,492	22,734	65,786	405	2,798	68,989				
1975 Household & Commercial	325	6,950	8,660	15,935	—	—	15,935	—	15,935	4,240	20,175
Industrial	4,600	6,510	11,740	22,850	—	—	22,850	—	22,850	3,010	25,860
Transportation	—	18,050	1,020	19,070	—	—	19,070	—	19,070	20	19,090
Electrical Generation	8,900	3,580	3,800	16,280	2,560	3,570	22,410	—	22,410	(7,270)	—
Synthetic Gas	—	—	—	—	—	—	—	—	—	—	—
Total	13,825	35,090	25,220	74,135	2,560	3,570	80,265				
1980 Household & Commercial	300	7,720	9,480	17,500	—	—	17,500	320	17,820	6,040	23,860
Industrial	4,750	7,590	12,500	24,840	—	—	24,840	380	25,220	4,170	29,390
Transportation	—	21,440	1,400	22,840	—	—	22,840	—	22,840	30	22,870
Electrical Generation	10,660	5,000	3,600	19,260	6,720	3,990	29,970	—	29,970	(10,240)	—
Synthetic Gas	430	440	—	870	—	—	870	(700)	—	—	—
Total	16,140	42,190	26,980	85,310	6,720	3,990	96,020				
1985 Household & Commercial	100	8,800	10,060	18,960	—	—	18,960	940	19,900	7,800	27,700
Industrial	5,150	9,130	13,240	27,520	—	—	27,520	1,060	28,580	6,290	34,870
Transportation	—	25,450	1,640	27,090	—	—	27,090	—	27,090	40	27,130
Electrical Generation	14,220	6,650	3,450	24,320	11,750	4,320	40,390	—	40,390	(14,130)	—
Synthetic Gas	2,000	670	—	2,670	—	—	2,670	(2,000)	—	—	—
Total	21,470	50,700	28,390	100,560	11,750	4,320	116,630				
2000 Household & Commercial	—	11,120	10,800	21,920	—	—	21,920	2,640	24,560	15,070	39,630
Industrial	6,700	14,660	17,940	39,300	—	—	39,300	2,860	42,160	15,620	57,780
Transportation	—	40,010	2,600	42,610	—	—	42,610	—	42,610	50	42,660
Electrical Generation	17,520	5,040	2,640	25,200	49,230	5,950	80,380	—	80,380	(30,740)	—
Synthetic Gas	7,140	550	—	7,690	—	—	7,690	(5,500)	—	—	—
Total	31,360	71,380	33,980	136,720	49,230	5,950	191,900				

[1] Includes anthracite, bituminous, and lignite.

[2] Petroleum products refined and processed from crude oil, including still gas, liquefied refinery gas, and natural gas liquids.

[3] Reproduced from "U.S. Energy Through the Year 2000", Walter G. Dupree, Jr., and James A. West, U.S. Dept. of the Interior (December 1972).

Table VI.1.

US AND FOREIGN ANNUAL U_3O_8 DEMAND*
(THOUSAND SHORT TONS)

	U.S.A.		Foreign Non-Communist		Foreign Communist	
EOCY	No Recycle (a)	Recycle (b)	No Recycle (a)	Recycle (c)	No Recycle (a)	Recycle (c)
1975	12.7	12.7	17.0	16.8	2.3	1.6
1976	15.2	15.2	20.1	19.2	2.6	1.9
1977	17.0	17.0	24.4	23.0	3.4	2.3
1978	21.4	21.1	30.2	27.8	4.5	2.9
1979	26.8	25.9	37.1	33.1	4.2	2.2
1980	30.7	29.6	47.8	42.5	6.0	3.7
1981	35.4	33.5	51.5	46.0	6.1	3.7
1982	42.3	38.6	58.8	52.4	7.4	4.4
1983	49.1	43.6	71.2	62.2	9.2	5.6
1984	54.0	47.6	80.9	69.2	10.9	6.8
1985	61.5	54.5	88.4	74.0	13.8	8.4
1986	69.4	60.7	102.2	85.6	17.0	10.5
1987	76.4	65.7	110.7	91.4	22.7	15.0
1988	83.5	70.4	125.1	104.0	27.3	18.4
1989	91.6	74.8	132.0	108.4	33.2	23.0
1990	100.8	78.1	147.2	120.5	39.1	27.2
1991	109.6	82.0	157.7	128.0	46.0	32.0
1992	117.6	86.4	170.0	142.4	53.3	38.8
1993	126.0	90.6	184.0	159.6	60.9	46.3
1994	133.6	93.1	195.7	170.3	68.6	53.5
1995	140.2	96.1	208.7	184.2	75.5	60.4
1996	146.9	108.0	223.2	199.1	81.5	66.1
1997	152.9	120.2	235.9	209.5	88.7	72.2
1998	157.1	123.1	249.1	218.5	94.2	76.1
1999	159.5	124.3	257.4	225.0	99.7	80.8
2000	163.8	127.3	272.5	238.0	105.1	85.4

*Tails Assay 0.3%; 72% Equilibrium Capacity Factor

(a) No Pu or U recycle

(b) Pu recycle beginning 1981 rising to 80% in 1987; U recycle beginning 1979; reprocessing capacity does not equal demand until 1996.

(c) Pu recycle beginning 1979 rising to 100% in 1982; U recycle begins 1974; reprocessing capacity equals reprocessing demand.

Source: EEI (Edison Electric Institute Nuclear Fuels Supply Study Program, 1976)

Table VI.2.

US AND FOREIGN CUMULATIVE U_3O_8 DEMAND*
(THOUSAND SHORT TONS)

	U.S.A.		Foreign Non-Communist		Foreign Communist	
EOCY	No Recycle (a)	Recycle (b)	No Recycle (a)	Recycle (c)	No Recycle (a)	Recycle (c)
1975	12.7	12.7	17.0	16.8	2.3	1.6
1976	27.9	27.9	37.1	36.0	4.9	3.5
1977	44.9	44.9	61.5	59.0	8.3	5.8
1978	66.3	65.9	91.7	86.8	12.8	8.7
1979	93.1	91.8	128.8	119.9	17.0	10.9
1980	123.8	121.4	176.6	162.4	23.0	14.6
1981	159.1	154.9	228.1	208.4	29.1	18.3
1982	201.4	193.6	286.9	260.8	36.5	22.7
1983	250.5	237.2	358.1	323.0	45.7	28.3
1984	304.5	284.8	439.0	392.2	56.6	35.1
1985	366.0	339.3	527.4	466.2	70.4	43.5
1986	435.4	400.0	629.6	551.8	87.4	54.0
1987	511.8	465.7	740.3	643.2	110.1	69.0
1988	595.3	536.1	865.4	747.2	137.4	87.4
1989	686.9	610.9	997.4	855.6	170.6	110.4
1990	787.7	689.0	1144.6	976.1	209.7	137.6
1991	897.2	771.0	1302.3	1104.1	255.7	169.6
1992	1014.9	857.4	1472.3	1246.5	309.0	208.4
1993	1140.9	947.9	1656.3	1406.1	369.9	254.7
1994	1274.5	1041.1	1852.0	1576.4	438.5	308.2
1995	1414.7	1137.2	2060.7	1760.6	514.0	368.6
1996	1561.6	1245.1	2283.9	1959.7	595.5	434.7
1997	1714.5	1365.3	2519.8	2169.2	684.2	506.9
1998	1871.5	1488.4	2768.9	2387.7	778.4	583.0
1999	2031.1	1612.7	3026.3	2612.7	878.1	663.8
2000	2194.8	1740.0	3298.8	2850.7	983.2	749.2

*Tails Assay 0.3%; 72% Equilibrium Capacity Factor

(a) No Pu or U recycle

(b) Pu recycle beginning 1981 rising to 80% in 1987; U recycle beginning 1979; reprocessing capacity does not equal demand until 1996.

(c) Pu recycle beginning 1979 rising to 100% in 1982; U recycle beginning 1974; reprocessing capacity equals reprocessing demand.

Source: EEI (Edison Electric Institute Nuclear Fuels Supply Study Program, 1976)

Table VI.3.

US AND NON-COMMUNIST FOREIGN NATURAL URANIUM RESOURCE POSITIONS (thousand STU_3O_8)

	Reserves (a)	Probable (b)	Total
U.S.A. (c, d):			
<$15/lb	430	655	1,090
<$30/lb	640 (e)	1,060	1,700
Foreign (f):			
<$15/lb	890	670	1,560
<$30/lb	1,590	1,080	2,670
Total:			
<$15/lb	1,320	1,325	2,650
<$30/lb	2,230	2,140	4,370

(a) Reasonably assured resources, i.e., reserves based on drilling.
(b) Estimated additional resources, i.e., probable.
(c) As reported in April 1976 by US ERDA.
(d) The US also has 1,750,000 STU_3O_8 at <$30/lb in "possible" and "speculative" categories.
(e) Does not include about 140,000 STU_3O_8 by-product of phosphate and copper production.
(f) As reported in late 1975.

Source: ERDA data compiled by Dr. Julian J. Steyn, of NUS Corp.

Table VI.4

US AND FOREIGN ANNUAL SWU DEMAND*
(MILLION SWU)

	U.S.A.		Foreign Non-Communist		Foreign Communist	
EOCY	No Recycle (a)	Recycle (b)	No Recycle (a)	Recycle (c)	No Recycle (a)	Recycle (c)
1975	4.6	4.6	4.3	4.3	.83	.66
1976	5.4	5.4	5.5	5.4	.82	.66
1977	6.5	6.5	5.9	5.7	1.1	.96
1978	6.8	6.8	8.5	8.0	1.2	.85
1979	10.0	9.9	11.3	10.5	1.8	1.3
1980	10.8	10.6	14.9	13.6	1.8	1.2
1981	12.8	12.5	17.0	15.1	2.3	1.7
1982	14.5	13.9	19.1	17.3	2.4	1.8
1983	18.2	17.4	22.0	19.6	2.9	2.2
1984	19.8	18.3	28.6	25.4	3.7	2.9
1985	22.3	21.1	32.2	28.2	4.1	3.2
1986	25.9	24.2	35.1	30.0	5.4	4.2
1987	28.6	26.5	37.9	31.8	6.6	5.2
1988	31.7	29.0	43.5	36.3	8.2	6.5
1989	34.5	31.0	46.7	38.6	10.2	8.3
1990	38.4	33.5	52.1	43.3	12.1	10.0
1991	42.1	35.2	55.6	46.0	14.2	11.7
1992	45.7	38.5	59.9	49.2	16.8	13.9
1993	49.0	40.9	65.1	57.3	19.2	16.9
1994	52.9	43.6	69.7	62.7	22.0	20.1
1995	55.4	45.1	73.9	66.8	24.4	22.4
1996	58.6	48.0	78.8	73.3	26.8	25.2
1997	61.1	53.9	83.0	77.1	29.1	27.6
1998	63.6	57.1	87.8	80.8	30.7	28.9
1999	64.9	57.2	91.2	82.8	32.5	30.3
2000	65.9	58.5	96.4	88.1	34.4	32.5

*Tails Assay 0.3%; 72% Equilibrium Capacity Factor
(a) No U or Pu recycle
(b) Pu recycle begins in 1981 and rises to 80% by 1987; U recycle begins in 1979; reprocessing capacity does not equal demand until 1996.
(c) U recycle begins 1974; Pu recycle begins in 1979 and rises to 100% by 1982; reprocessing capacity equals reprocessing demand.

Source: EEI (Edison Electric Institute Nuclear Fuels Supply Study Program, 1976)

Table VI.5

US AND FOREIGN CUMULATIVE SWU DEMAND*
(MILLION SWU)

	U.S.A.		Foreign Non-Communist		Foreign Communist	
EOCY	No Recycle (a)	Recycle (b)	No Recycle (a)	Recycle (c)	No Recycle (a)	Recycle (c)
1975	4.6	4.6	4.3	4.3	.83	.66
1976	9.9	9.9	9.8	9.7	1.6	1.3
1977	16.5	16.5	15.7	15.4	2.7	2.3
1978	23.2	23.2	24.2	23.4	3.9	3.2
1979	33.2	33.1	35.5	33.9	5.7	4.5
1980	44.0	43.8	50.4	47.5	7.5	5.7
1981	56.8	56.3	67.4	62.6	9.8	7.4
1982	71.3	70.2	86.5	79.9	12.2	9.2
1983	89.5	87.7	108.5	99.5	15.1	11.4
1984	109.3	106.0	137.1	124.9	18.8	14.3
1985	131.6	127.1	169.3	153.1	22.9	17.5
1986	157.6	151.3	204.4	183.1	28.3	21.7
1987	186.2	177.8	242.3	214.9	34.9	26.9
1988	217.9	206.8	285.8	251.2	43.1	33.4
1989	252.5	237.7	332.5	289.8	53.3	41.7
1990	290.9	271.2	384.6	333.1	65.4	51.7
1991	333.0	306.4	440.2	379.1	79.6	63.4
1992	378.8	344.9	500.1	428.3	96.4	77.3
1993	427.8	385.8	565.2	485.6	115.6	94.2
1994	480.6	429.4	634.9	548.3	139.6	114.3
1995	536.1	474.4	708.8	615.1	162.0	136.7
1996	594.7	522.4	787.6	688.4	188.8	161.9
1997	655.8	576.3	870.6	765.5	217.9	189.5
1998	719.4	633.4	958.4	846.3	248.6	218.4
1999	784.3	690.6	1049.6	929.1	281.1	248.7
2000	850.2	749.1	1146.0	1017.0	315.5	281.2

*Tails Assay 0.3%; 72% Equilibrium Capacity Factor
(a) No U or Pu recycle
(b) Pu recycle begins in 1981 and rises to 80% by 1987; U recycle begins in 1979; reprocessing capacity does not equal demand until 1996.
(c) U recycle begins 1974; Pu recycle begins in 1979 and rises to 100% by 1982; reprocessing capacity equals reprocessing demand.

Source: EEI (Edison Electric Institute Nuclear Fuels Supply Study Program, 1976)

Table VI.6.

PROJECTED DISCHARGE OF U.S. REACTOR FUEL (Metric Tons Heavy Metal)

End of Calendar Year	Annual	Cumulative
1975	612	1,568 (a)
1976	928	2,495
1977	1,133	3,629
1978	1,316	4,948
1979	1,602	6,549
1980	1,882	8,431
1981	2,137	10,568
1982	2,482	13,050
1983	2,994	16,044
1984	3,500	19,544
1985	4,111	23,655
1986	4,875	28,530
1987	5,452	33,982
1988	6,096	40,078
1989	6,835	46,913
1990	7,556	54,469
1991	8,315	62,784
1992	9,076	71,860
1993	9,900	81,760
1994	10,795	92,555
1995	11,696	104,251
1996	12,560	116,811
1997	13,389	130,200
1998	14,144	144,344
1999	14,819	159,163
2000	15,439	174,602

(a) Includes 955 MTHM inventory at 12/31/74

Source: EEI (Edison Electric Institute Nuclear Fuels Supply Study Program, 1976)

Table VI.7.

U.S. SPENT FUEL REPROCESSING DEMAND (a)
(Metric Tons Heavy Metal)

End of Calendar Year	Annual	Cumulative
1975	422	422 (b)
1976	803	1,225
1977	1,052	2,277
1978	1,214	3,491
1979	1,418	4,909
1980	1,786	6,695
1981	1,978	8,673
1982	2,297	10,970
1983	2,668	13,638
1984	3,319	16,957
1985	3,682	20,639
1986	4,539	25,178
1987	5,212	30,390
1988	5,693	36,083
1989	6,498	42,581
1990	7,172	49,753
1991	7,941	57,694
1992	8,689	66,383
1993	9,463	75,846
1994	10,336	86,182
1995	11,254	97,436
1996	12,138	109,574
1997	12,987	122,561
1998	13,791	136,352
1999	14,497	150,849
2000	15,141	165,990

(a) Light water reactors only
(b) Does not include 955 MTHM in inventory as at 12/31/74.

Source: EEI (Edison Electric Institute Nuclear Fuels Supply Study Program, 1976)

Table VI.8.

ESTIMATED WORLD NATURAL URANIUM PRODUCTION CAPACITIES (a)
(Tons U_3O_8)

Country	Current (b)	Attainable 1978
Argentina	210	670
Australia	—	2,600 (c)
Canada	8,450	11,050 (d)
France	2,340	2,860
Gabon	780	1,560
Germany	320	320
Japan	40	40
Mexico	—	320
Niger	1,950	1,950
Portugal	120	140
Spain	190	440
South Africa	4,940	14,300 (e)
USA	18,000	24,700 (f)
Yugoslavia	—	300
Total	37,340	61,250

(a) Reference pending publication.
(b) 1974 production from Canada, France, Germany, Japan, Portugal, Spain and USA totalled ~18,220 STU_3O_8 or 69% of nominal plant capacity; 1975 projected to be 70.5%.
(c) May be too optimistic for either 1978 or 1979 (see *Appendix C*).
(d) 10,800 STU_3O_8/year per H. B. Merlin, Canadian Government.
(e) South African production tied largely to gold mining and, thus, tied to gold market price, i.e., uranium productivity low when gold price low.
(f) 24,100 STU_3O_8/year per J. F. Facer, ERDA Grand Junction Office, October 1975.

Source: IEA/OECD data

Table VI.9.

PROJECTED NUMBER OF NEW NATURAL URANIUM MINE/MILLS REQUIRED TO MEET PROJECTED NON-COMMUNIST WORLD DEMAND (CAPACITY OF 1,000 STU_3O_8/YEAR AND 80% CAPACITY FACTOR ASSUMED) (a)

	Non-Comm. World (b, d)		USA (c, d)	
	No Recycle	Recycle	No Recycle	Recycle
1982	58	45	25	20
1985	119	92	49	41
1990	241	180	99	70
2000	477	388	177	131

(a) 0.3% tails assay and 72% CF for nuclear plants; 80% CF for mine/mills.
(b) 49,000 STU_3U_8/year (existing but expanded actual capacity) assumed by 1978.
(c) 20,000 STU_3O_8/year (existing but expanded actual capacity) assumed by 1978.
(d) 25,000 STU_3O_8 US inventory and 48,000 STU_3O_8 foreign inventory assumed at 1/1/75; 1980 existing capacity taken as 22,000 STU_3O_8/year for US and 55,000 STU_3O_8/year for non-Communist world.

Source: EEI (Edison Electric Institute Nuclear Fuels Supply Study Program, 1976)

Table VI.10.

SUMMARY STATUS OF NON-COMMUNIST WORLD U_3O_8 TO UF_6 CONVERSION SUPPLY CAPACITY

Country	Plant	Annual Capacity (tons uranium) Current 1975	Annual Capacity (tons uranium) Planned 1978
USA	Allied Chemical (Metropolis, Illinois)	14,000	14,000
	Kerr-McGee Nuclear (Sesquoiah, Oklahoma)	5,000	10,000
Canada	Eldorado Nuclear Limited (Port Hope, Ontario)	4,550	6,800
United Kingdom	British Nuclear Fuels Ltd. (Springfield, Lancashire)	5,000	8,000
France	COMHUREX (a) (Pierrelatte, South France)	7,500	10,000
	Total	36,000	48,800

(a) Société pour la Conversion de l' Uranium en Métal et en Hexafluoride.

Source: EEI (Edison Electric Institute Nuclear Fuels Supply Study Program, 1976)

Table VI.11.

PROJECTED SCHEDULE OF WORLD SWU SUPPLY FROM ENRICHMENT PLANTS CURRENTLY IN OPERATION (million SWU)

	Existing Supply to Non-Communist World					Non-Communist World Demand
	Annual					
Calendar Year	ERDA (a)	USSR (b)	France & UK (c)	Total	Cum Total	Cum Total (d)
1975	13.6	.4	.6	14.6	32.6 (e)	8.9
1976	15.0	.8	.6	16.4	49.0	19.6
1977	14.9	1.6	.6	17.1	66.1	32.2
1978	14.7	2.5	.6	17.8	83.9	47.4
1979	15.7	3.0	.6	19.3	103.2	68.7
1980	16.8	3.0	.6	20.4	123.6	94.5
1981	16.8	3.0	.5	20.3	143.9	124.2
1982	16.8	3.0	.3	20.1	164.0	157.8
1983	16.8	3.0	—	19.8	183.8	198.0
1984	17.1	3.0	—	20.1	203.9	246.4
1985	17.2	3.0	—	20.2	224.1	300.9
1986	17.2	3.0	—	20.2	244.3	362.0
1987	17.2	3.0	—	20.2	264.5	428.5

(a) Existing three-site diffusion plant complex, excludes CIP and CUP.
(b) Estimated USSR export fraction; existing nominal capacity believed to be ~8 million SWU/year.
(c) Existing Pierrelatte and Capenhurst diffusion plants, assumed to be phased out by 1982.
(d) Moderate growth case assuming no U or Pu recycle, 0.3% tails assay, 72% equilibrium capacity factor; excludes Communist nations (except Yugoslavia).
(e) Includes ERDA inventory of 18 million SWU at 1/1/75.

Source: EEI (Edison Electric Institute Nuclear Fuels Supply Study Program, 1976)

Table VI.12.

PROJECTED SCHEDULE OF WORLD SWU SUPPLY FROM ENRICHMENT CAPACITY PLANNED & AUTHORIZED (million SWU)

	Planned & Authorized Supply					Operating & Planned Supply (d)	Non-Communist World Demand (e)
Calendar	Annual						
Year	ERDA (a)	TRICASTIN (b)	URENCO (c)	Total	Cumulative Total	Cumulative Total	Cumulative Total
1975	—	—	—	—	—	32.6 (f)	8.9
1976	1.1	—	—	1.1	1.1	50.1	19.6
1977	2.2	—	—	2.2	3.3	69.4	32.2
1978	3.7	—	0.2	3.9	7.2	91.1	47.4
1979	5.9	1.5	0.5	7.9	15.1	118.3	68.7
1980	7.8	4.5	1.0	13.3	28.4	152.0	94.5
1981	8.5	7.5	1.4	17.4	45.8	189.7	124.2
1982	8.5	10.2	2.0	20.7	66.5	230.5	157.8
1983	8.7	10.7	2.0	21.4	87.9	271.7	198.0
1984	9.6	10.7	2.0	22.3	110.2	314.1	246.4
1985	10.5	10.7	2.0	23.2	133.4	357.5	300.9
1986	10.5	10.7	2.0	23.2	156.6	400.9	362.0
1987	10.5	10.7	2.0	23.2	179.8	444.3	428.5
1988	10.5	10.7	2.0	23.2	203.0	487.7	503.7

(a) CIP and CUP; planned, authorized in part and in early stages of addition.
(b) Eurodif planned and authorized diffusion plant; in early stages of construction.
(c) Capacity of ∼0.5 million SWU/year under construction; contract commitments for 2 million SWU/year obtained; further expansion dependent upon Urenco marketing success.
(d) Includes plants in operation from Table VI.11; i.e., represents 'firm' supply while excluding 'potential'.
(e) Excludes US Government requirements, stockpiles, working inventories, new plant flywheel and similar items; moderate growth assuming no Pu or U recycle, 0.3% tails assay and 72% equilibrium capacity factor.
(f) Includes ERDA inventory of 18 million SWU as at 1/1/75.

Source: EEI (Edison Electric Institute Nuclear Fuels Supply Study Program, 1976)

Table VI.13.

NEW 6 MILLION SWU/YEAR ENRICHMENT CAPACITY ADDITIONS TO MEET NON-COMMUNIST WORLD SHORTFALL (a) (Million SWU)

End Calendar Year	No. of Plants (b)		Capacity		Required (c)
	Started Up	Fully Operating	Annual	Cumulative	Cumulative
1984	1	0	3	3	(38)
1985	1	1	6	9	(23)
1986	2	1	9	18	1
1987	4	2	18	36	27
1988	6	4	30	66	64
1989	8	6	42	108	106
1990	9	8	51	159	157
1991	10	9	57	216	216
1992	12	10	66	282	282
1993	13	12	75	357	357
1994	15	13	84	441	441
1995	15	15	90	531	531
1996	18	15	99	630	629
1997	18	18	108	738	734
1998	19	18	111	849	847
1999	20	19	117	966	964
2000	20	20	120	1,086	1,086

(a) Based on 'no recycle'; does not include new plant flywheel or US Government requirements.
(b) Assumes 3 million SWU in start-up year; e.g., annual capacity for year 1992 = $(12 - 10) \times 3 + 10 \times 66 = 66$ million SWU.
(c) Stockpile quantities shown in parentheses.

Source: EEI (Edison Electric Institute Nuclear Fuels Supply Study Program, 1976)

Table VI.14.

REPROCESSING PLANT SCHEDULE AND REQUIREMENTS TO MEET REPROCESSING DEMAND

End of Calendar Year	Total (a) No. of Reprocessing Plants	Cumulative (b) No. of New Plants Started	Annual New Plant Capacity MTHM/yr	Net Annual Capacity MTHM/yr	Total Cumulative Reprocessing Through-put	Cumulative (c) Storage at Reprocessing Plant MTHM	Total Cumulative Demand for Reprocessing	Outside Storage or Reprocessing Shortfall
1984	2	0	0	2,250	7,125	1,750	17,912	9,037
1985	2	0	0	2,250	9,375	1,750	21,594	10,469
1986	3	1	600	2,850	12,225	2,110	26,133	11,798
1987	3	1	1,500	3,750	15,975	2,470	31,345	12,900
1988	4	2	2,100	4,350	20,325	2,830	37,038	13,883
1989	5	3	3,600	5,850	26,175	3,190	43,536	14,171
1990	6	4	5,100	7,350	33,525	3,550	50,708	13,633
1991	7	5	6,600	8,850	42,375	3,910	58,649	12,364
1992	8	6	8,100	10,350	52,725	4,270	67,338	10,343
1993	9	7	9,600	11,850	64,575	4,630	76,801	7,596
1994	10	8	11,100	13,350	77,925	4,630	87,137	4,582
1995	11	9	12,600	14,850	92,775	4,630	98,391	986
1996	11	9	13,500	15,750	108,525	4,630	110,529	
1997	11	9	13,500	15,750	124,275	4,630	123,516	
1998	11	9	13,500	15,750	140,025	4,630	137,307	
1999	11	9	13,500	15,750	155,775	4,630	151,804	
2000	11	9	13,500	15,750	171,525	4,630	166,945	

(a) Assumes Barnwell, S.C., and West Valley, N.Y., reprocessing plant start-ups in 1979 and 1982 respectively.
(b) Each new reprocessing plant is assumed to be 1500 MTHM/year capacity; 600 and 1500 MTHM are assumed in first and second years.
(c) Each new reprocessing plant adds 360 MTHM storage capacity 2 years ahead of reprocessing startup.

Source: EEI (Edison Electric Institute Nuclear Fuels Supply Study Program, 1976)

Figure I.1.

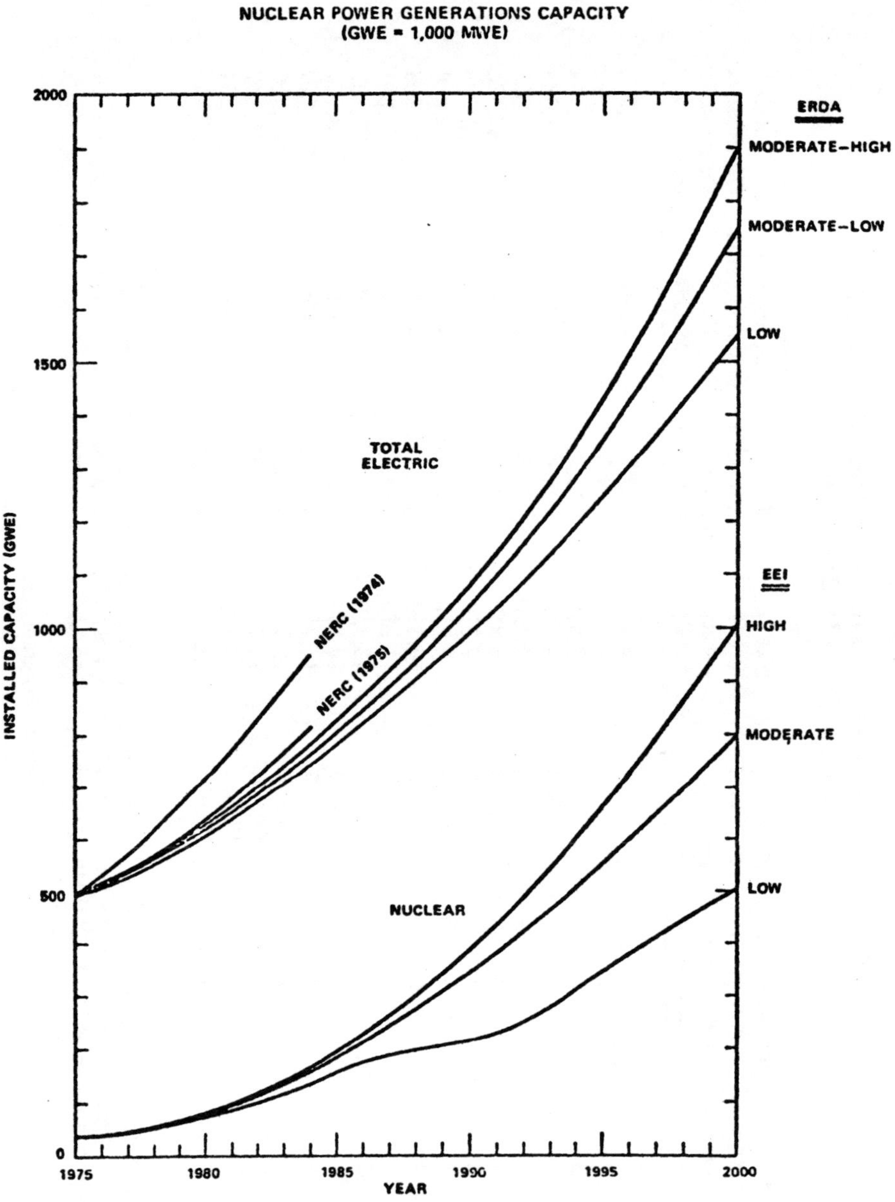

Source: EEI (Edison Electric Institute Nuclear Fuels Supply Study Program, 1976)

Figure VI.1.

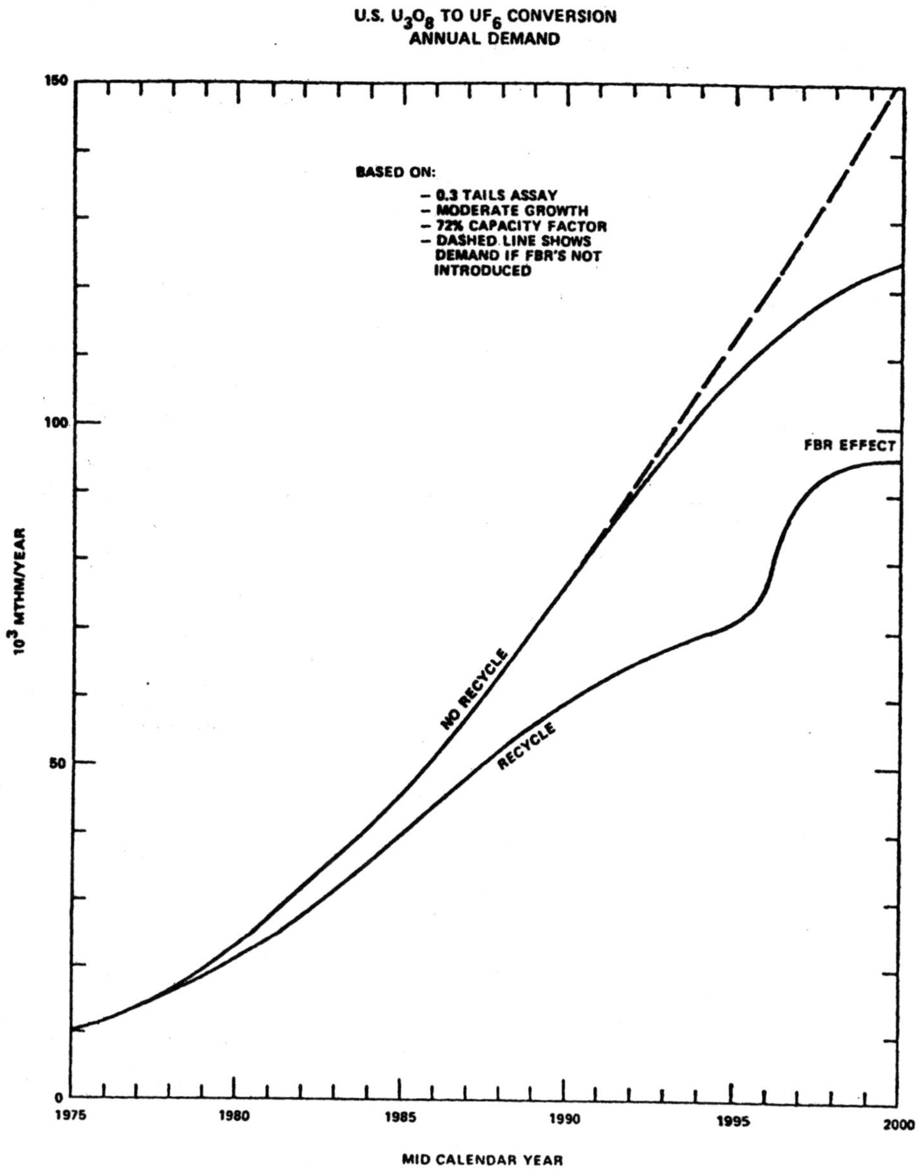

Source: EEI (Edison Electric Institute Nuclear Fuels Supply Study Program, 1976)

Figure VI.2.

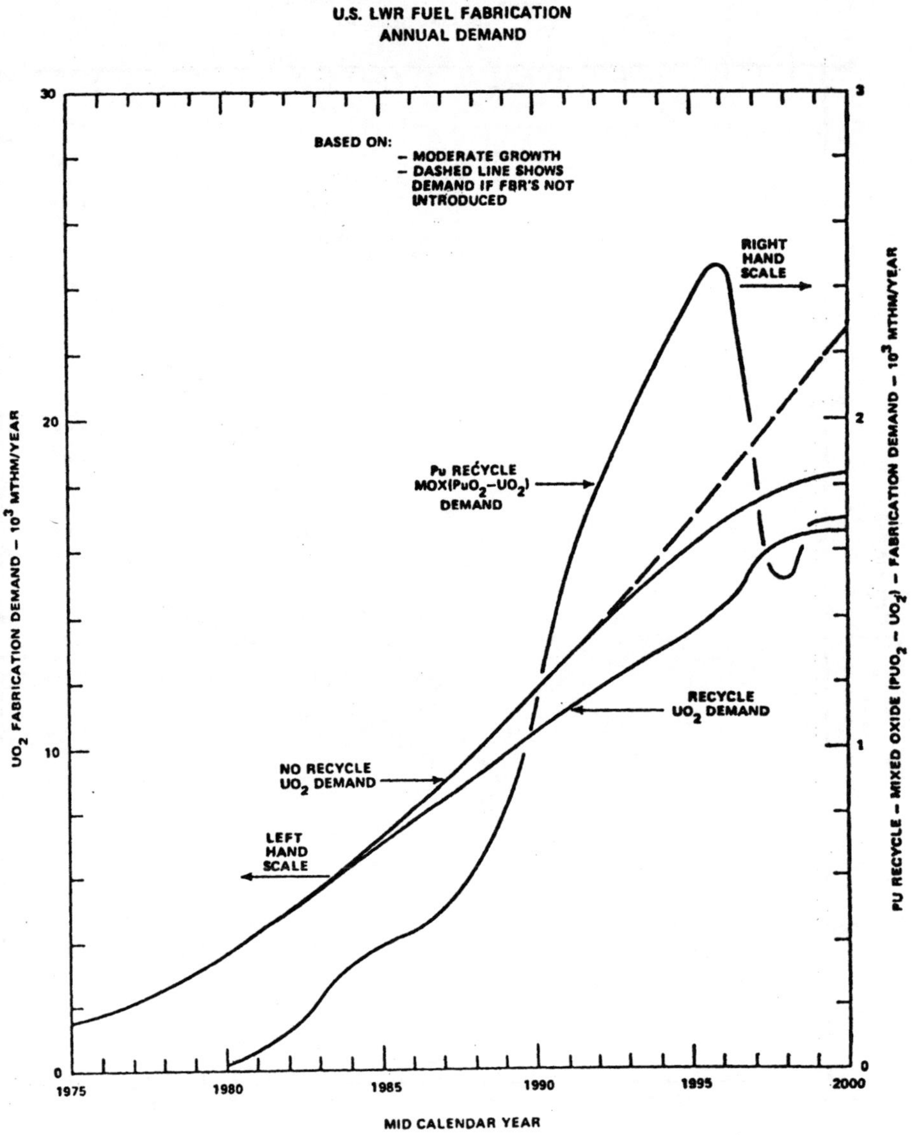

Source: EEI (Edison Electric Institute Nuclear Fuels Supply Study Program, 1976)

Figure VI.3.

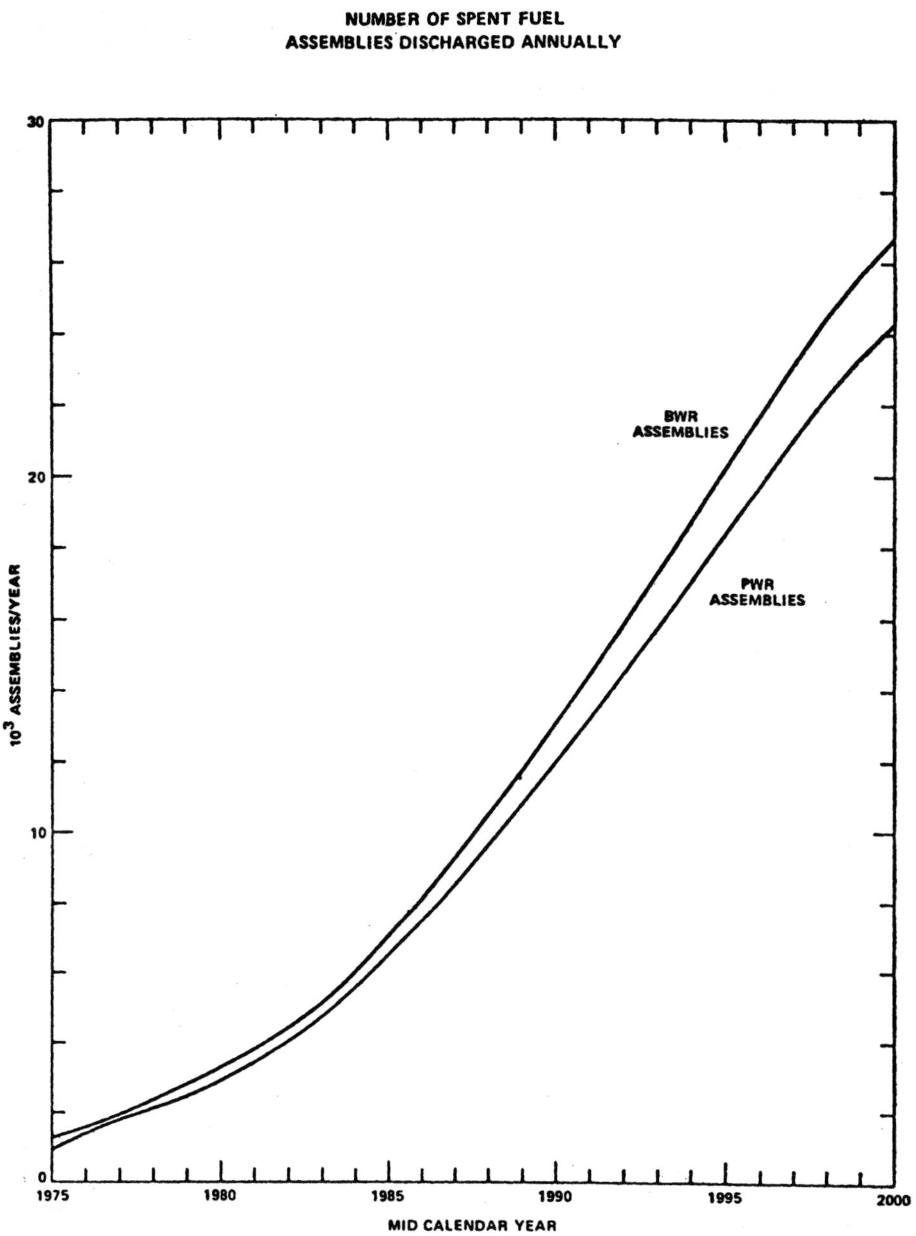

Source: EEI (Edison Electric Institute Nuclear Fuels Supply Study Program, 1976)

Figure VI.4.

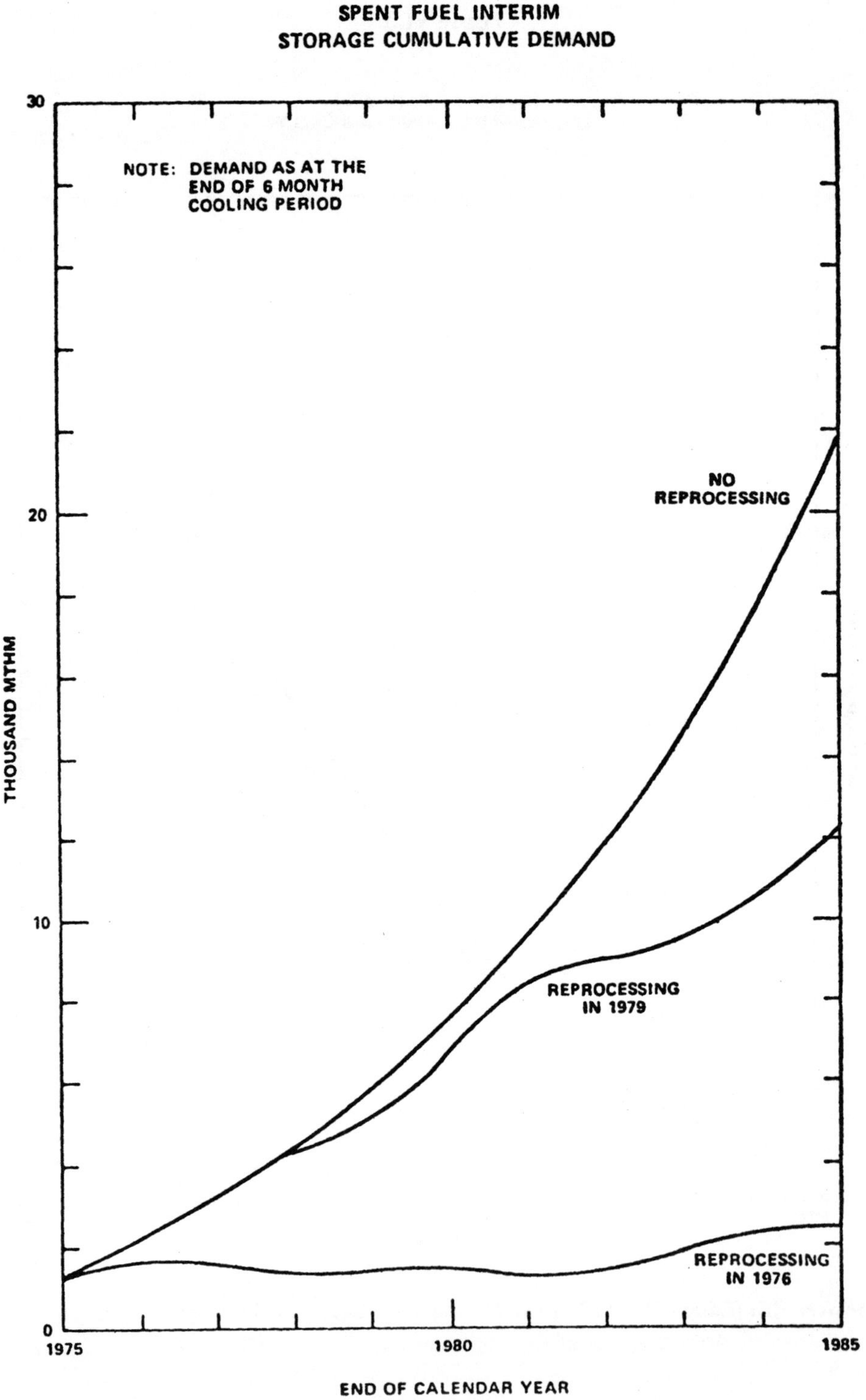

Source: EEI (Edison Electric Institute Nuclear Fuels Supply Study Program, 1976)

APPENDIX B

THE NUCLEAR FUEL CYCLE

I. An Historical Perspective

The nuclear fuel cycle, and all operations and problems attendant to it, have been around since the war years of the early 1940's. In order to produce atomic weapons from scratch, two paths were followed. One was to enrich uranium to "weapons grade" concentrations of U^{235}. The other was to fabricate the natural uranium into fuel elements which were then processed through "production reactors" in which plutonium was created by neutron irradiation. The product—plutonium—was separated from the uranium in chemical reprocessing plants and made into components for use in atomic weapons. The wastes from the reprocessing plants containing highly radioactive fission products were and still are stored as liquids in large steel tanks.

Uranium needed for both processes was mined from known deposits in Africa, the US and Canada, and from others subsequently discovered. The industrial-type processing operations were carried out at vast installations located in isolated regions of the US. Major US industries developed, designed, built and operated these facilities under the direction first of the US Army Corps of Engineers and later of the US Atomic Energy Commission. All of these nuclear fuel cycle operations which are related to weapons production were and still are carried out under tight government security regulations.

These war-time operations were the origins from which emerged organizations, technical personnel, technology and experience which provided the initial base for the US naval nuclear propulsion program and the US civilian reactor program. The nuclear fuel cycle complex, however, remained under the control of the US AEC. This fact resulted in providing the nuclear industry not only relatively early access to, but a strong dependence upon US government facilities and organizations comfortably obscured from public view, and a vast body of technology that had been developed out of military need. The Congress, through a dedicated and paternalistic Joint Committee on Atomic Energy, monitored and supported a rather effective use of resources of manpower and facilities needed to satisfy military requirements while at the same time holding forth the promise of a new source of power for peaceful applications.

Following World War II the various nations of the world went their separate ways in the pursuit of national objectives derived from the demonstrated release of vast amounts of energy from the atomic nucleus. The United Kingdom, France and the USSR all established programs for the production of weapons based upon both enriched uranium and plutonium. All three countries also mounted programs for the production of electricity from nuclear energy.

Japan, West Germany, Italy and Sweden all eschewed the pursuit of military weapons programs but have subsequently mounted significant efforts to engage their industry in producing plants and equipment for the application of nuclear energy to peaceful uses.

II. The Dynamics of the Supply Equation: A Layman's Guide to the Nuclear Fuel Cycle

Nuclear Fuel Cycle Summary

- Uranium exploration, mining and milling.
- Uranium hexafluoride production (an interim chemical conversion to transform uranium into a gaseous compound in preparation for enrichment).
- Isotope enrichment (whereby the "separative work" of concentrating the necessary lighter isotope of uranium, U^{235}, is performed currently).
- Fuel fabrication (a "manufacturing" step that forms the uranium enriched in U^{235}, the isotope that fissions in the reactor, into pellets placed in long tubes or rods that form the reactor core).
- Fuel irradiation in the nuclear power reactor (a heat source, where the U^{235} fissions, producing heat used to make the steam that drives turbines which, in turn, power electric generators).
- Fuel reprocessing and recycling (whereby spent uranium fuel is treated as a "mineable resource" from which plutonium and "unburnt" uranium may be extracted as potential secondary fuel supplies).
- Plutonium storage (which safeguards the plutonium until it is used as fuel or otherwise disposed of).
- Federal and commercial waste management (to contain, store and safeguard for indefinite periods of time the highly radioactive portion of the spent fuel not recycled, principally the fission products).
- While the above concrete steps are set out diagrammatically below, the following "intangible" steps must also be considered as an integral part of the fuel cycle:
  - —financing
  - —safeguards
  - —insurance
  - —transport
  - —assaying
  - —inspection
  - —brokerage

NUCLEAR FUEL CYCLE

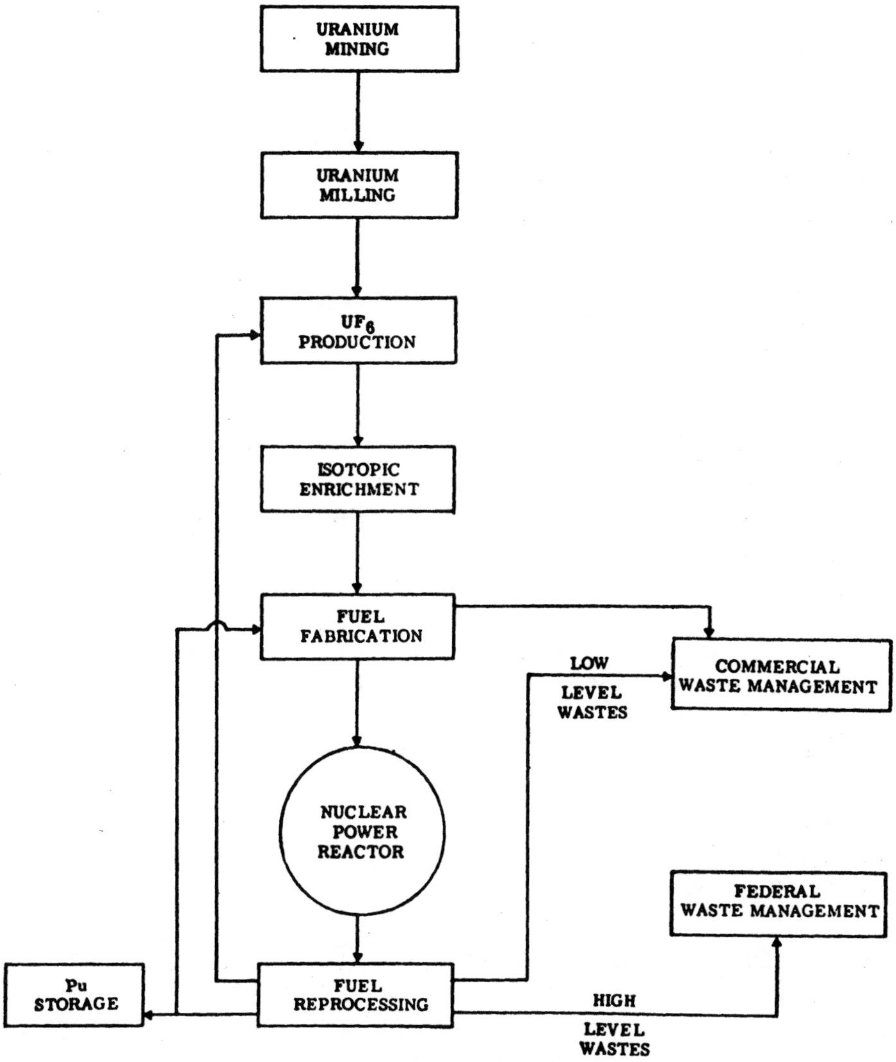

A Guide to the Nuclear Fuel Cycle

The nuclear fuel cycle consists of all those processes which are required for the economic utilization of nuclear fuel. These processes can be separated into three categories:

1. Front end
2. Irradiation
3. Tail end

The *front end of the fuel cycle* consists of those activities required to produce fuel assemblies suitable for use in nuclear power systems. The front end includes:

1. Mining and milling of natural uranium;
2. Conversion of uranium mill product to uranium hexafluoride (UF_6);
3. Enrichment of uranium;
4. Fabrication of fuel assemblies.

The *irradiation of nuclear fuel* is the step during which power is extracted from the fuel through the fission process. In addition to the production of power, through the utilization of uranium, plutonium and fission products are also produced.

The *tail end of the fuel cycle* consists of those processes and activities required to dispose of the radioactive waste produced during the irradiation step and to reclaim the remaining uranium and produced plutonium from the discharged fuel so that they may be utilized as feed in the front end of the fuel cycle. The tail end includes:

1. Temporary fuel storage;
2. Transportation of spent fuel;
3. Reprocessing of spent fuel;
4. Waste storage.

The nuclear fuel cycle is a dynamic system. That is, there is not a constant relationship between the amount of uranium mined to uranium product charged into a nuclear power plant. The variability of such key parameters on the amount of uranium feed required for a given amount of fabricated fuel results from flexibility inherent in the enrichment process. Additionally, the requirements of the front end of the fuel cycle for uranium and enrichment are affected by the availability and recovery efficiency of uranium and plutonium from the tail end. Therefore, in order to fully understand the tradeoffs and options available in the supply capability for nuclear fuel, one must have an understanding of the processes, costs, and timing of the various portions of the fuel cycle.

In order to put the material quantity requirements into perspective, the following information relevant to a single 1000 MWe light water reactor is of interest. The initial *enriched* uranium loading of the reactor core is approximately 100,000 kilograms; approximately 35,000 kilograms of enriched uranium are required annually during the life of the reactor. The *natural* uranium required to produce the enriched uranium utilized in the plant is approximately 1 million pounds for the initial core with an annual makeup of approximately 400,000 pounds. Thus over 30 years, a 1000 MWe nuclear plant will utilize approximately 12 million pounds of uranium.

Uranium Mining and Milling

Uranium is found as a naturally occurring mineral in various forms in the earth's crust. The quantity of uranium in the ore in which it is found is not usually great. Concentrations of 2-5 pounds of uranium per ton of ore are common in commercial grade ores, while concentrations as high as 50 pounds per ton ore are considered very rich and are unusual. Uranium ores are located both in deep deposits which have to be mined through conventional underground means and in deposits near to the surface which can be mined as open pits. The equipment and men utilized for the development and production of uranium deposits are quite similar to those utilized in the coal industry. Once mined, uranium ore is processed in a mill which uses a number of conventional, mechanical and chemical steps to separate the uranium from the host rock and other minerals. The uranium is recovered as yellowcake, a uranium salt containing between 70 and 80 percent U_3O_8, the shorthand notation for the product formula.

The availability of raw material or yellowcake for the nuclear fuel cycle is dependent upon the quantity of uranium in defined reserves, the installation of production or mining capacity to produce these reserves, and the construction of milling capability to process the ore into a usable raw material.* The size of the reserve base is a function of production cost and, therefore, a function of the market price for the product.

The development of a uranium mining and milling complex takes approximately ten years from initial exploration to full production. A facility capable of producing enough uranium to support six nuclear power plants will cost from $33 to 129 million. Additional capital charges of from $33 to 129 million would be required over the life of the facility, resulting in a total investment per reactor supported of from $11 to 42 million or from $11 to 43 per kw depending upon ore grade and whether or not there is recycle.

---

*In uranium, as in copper and other minerals which do not occur in discrete high grade deposits as does for example petroleum, the quantity of resources available is a function of the price at which recovery can be accomplished. For example, given a constant cost of mining of $30 per ton ore, reserves that contain three pounds of uranium per ton ore will cost $10 per pound of uranium product, while reserves that contain 0.3 pounds of uranium per ton ore will cost $100 per pound of uranium product. Recent ERDA estimates, which indicate an increase in cost from $10 per pound to $30 per pound, increase the domestic reserve base from 230,000 tons U_3O_8 to 600,000 tons U_3O_8 and the probable reserves from 440,000 tons to 1,100,000 tons.

Conversion of Uranium

The next step in the fuel cycle is the conversion of yellowcake into a form suitable for the enriching process. The form is a gas called uranium hexafluoride (UF_6). The process utilized is the fluoride volatility process involving three steps and concluding with a high temperature fluoridation step.

The UF_6 product is solid at room temperature but sublimes at approximately 134° F. The current cost of uranium conversion is approximately $1.25 per pound. A conversion plant capable of supporting 30 power plants is estimated to cost between $27 and 49 million and will take four years to design and build. The investment per plant is, therefore, $0.9 to 1.3 million or $0.9 to 1.3 per kw depending upon whether or not there is recycle.

Enrichment

Uranium, as found in nature, consists of several isotopes; the two principal ones being uranium 235 (U^{235}) and uranium 238 (U^{238}). Of these two, it is the uranium 235 which is readily fissible. The concentrations of naturally occurring U^{235} is approximately 0.7%. In order to utilize uranium efficiently and economically in light water reactors*, the concentration of U^{235} must be raised from 0.7% to about 3%. In order to utilize uranium in nuclear weapons, the U^{235} concentration must be raised to about 90%. The process of raising the concentration of U^{235} in uranium is called enrichment.

Because U^{235} and U^{238} are the same chemical element, the enrichment process must use differences in physical properties of the isotopes rather than chemical properties to obtain the desired increase in U^{235} concentration.

The most well developed and widely used method of enrichment is the gaseous diffusion process. This process makes use of the differences in rates at which gases or vapors of different molecular weight diffuse through a porous barrier.

The other method of enrichment currently planned for wide use is the gas centrifuge method. The principle of centrifuge enrichment is based upon the variation of the centrifugal force acting on a gas molecule with the molecular weight of the gas.

Other methods of enrichment currently under consideration include the laser and nozzle methods.

The effort involved in any of these enrichment techniques is expressed in terms of separative work units (SWU).**

*There are other reactor systems, notably the Canadian CANDU system which utilizes natural uranium as fuel and requires no enrichment.

**In order to produce 1 kilogram of 2.8% U^{235} product (.280 kg U^{235}/1 kg U = 2.8%), 6 kilograms of natural uranium feed (.0427 kg U^{235}/6 kg U = .71%) must be supplied to the

The current cost of enrichment at US ERDA facilities is between \$3 and 61 per SWU (depending on contract type). The "commercial" cost for services from the government facilities has been estimated by ERDA at \$76 per SWU.

The amount of uranium removed from the tails stream affects both the amount of SWU required for the enrichment process and the amount of uranium feed required.** Thus, by lowering the concentration of U^{235} in the tails stream, one is more effectively utilizing feed material as well as requiring more work from the enrichment plant. Conversely, by raising the concentration of U^{235} in the tails stream, one is less effectively utilizing feed uranium and requiring less work from the enrichment plant.

The development of a uranium enrichment complex requires approximately eight years. A facility (either gaseous diffusion or centrifuge) capable of supporting 120 power plants is estimated to cost \$3.3 billion. (In the case of gaseous diffusion, this does not include the cost of supporting power facilities, estimated at \$2 billion.) The required investment per plant is, therefore, \$28 million or \$28 per kw. In the event that added enrichment capacity is required to offset lack of recycle the investment per plant is \$37.5 million or \$37.5 per kw.

---

enrichment plant and three SWU are utilized. The process also results in the production of five kilograms of 0.3% U^{235} (.0147 kg U^{235}/5 kg U = 0.3%) waste or tails.

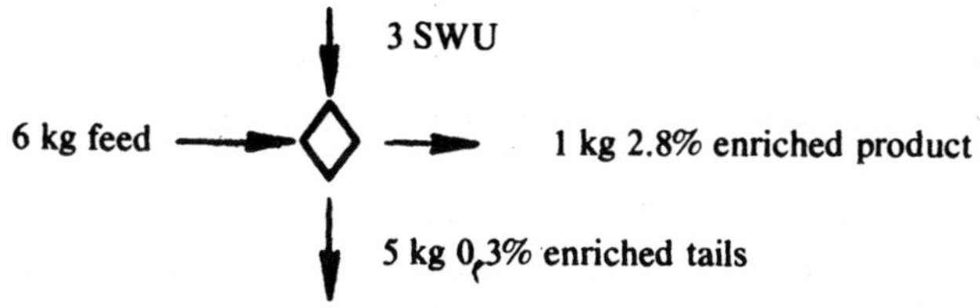

Thus, what has happened in the enrichment process is that .0280 kg of the 0.427 kg U^{235} originally present in 6 kilograms of feed has been concentrated in 1 kilogram of product. The remaining .0147 kg U^{235} (.0427 − .0280 = .0147) is discharged in 5 kilograms of 0.3% waste.

**The feed/product/tails/SWU relationship for a case where the concentration of U^{235} in tails is 0.4%. It is seen that this combination requires 7.7 kg of feed (in contrast to 6 @ 0.3% tails to produce 1 kg of 2.8% enriched product while utilizing 2.5 SWU (in contrast to 3 @ 0.3% tails).

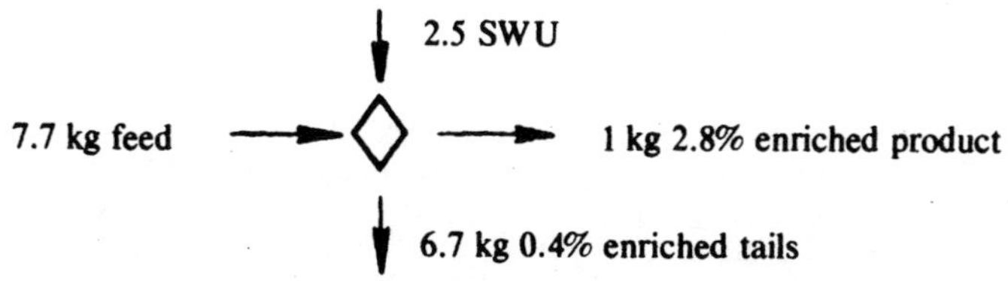

There is a strong relationship between uranium feed requirements, separative work requirements, and tails assay. It is seen from the above that a change in tails assay from 0.3% to 0.4% results in an increased feed requirement of 28% and a decreased SWU requirement of 17%.

Fabrication

The final step in the front end of the fuel cycle is fabrication. The enriched uranium produced in the enrichment plant is converted into small ceramic pellets of uranium dioxide. These pellets are then inserted into an assembly structure. When this is completed, the assembly is ready for shipment to the nuclear power plant for irradiation. The current cost of fabrication is $80 per kg uranium.

The construction of a fuel fabrication facility required to support 50 reactors (1,500,000 kg fuel per year) will take five years to construct and is estimated to cost from $90 to 155 million depending on the status of recycle. Therefore, the investment per plant is $1.1 to 3.1 million or $1.8 to 3.1 per kw.

Irradiation

The production of power from the nuclear fuel fabricated in the front end of the fuel cycle occurs during the irradiation step of the fuel cycle. There are two important processes which occur in the fuel during irradiation: fission, and capture of neutrons in U^{238}. It is in the fission process that energy is released.*

In addition, waste fission products which, after their separation from the fuel in reprocessing, result in the need for long-term waste storage.

The capture of neutrons in U^{238} without the occurrence of fission results in the production of plutonium. Plutonium, as discussed below, can be used as a substitute for U^{235} as a fissile material and utilized in feed material for new fuel.

Irradiation in nuclear plants occurs over a 3-4 year period once the plant is in equilibrium conditions. During this time, each kilogram of fuel will produce approximately 2000 million BTU resulting in the generation of 200,000 KWHe.

When the fuel is ready for discharge after its irradiation, the initial enrichment of 2.8% will be reduced to approximately that of natural uranium (7 gms U^{235}/kg U). The fuel will also contain approximately five grams of plutonium per kilogram of uranium. Twenty-five grams of fission products will also be contained.

Spent Fuel Storage

After the fuel has been irradiated and can no longer be effectively utilized, the fuel is removed from the core of the nuclear plant and stored for several

*Each fission event produces on the average about 200 million electron volts or 5×10^{-25} kw hrs. A fission rate of 5×10^{18} per second is required to produce 1 watt. Despite the relatively large number of fission events occurring, the power production from a 1000 MWe power plant is obtained from the utilization of about four pounds of U^{235} per day. (To obtain the same power through combustion, one requires 8000 tons of coal.)

The two other products of the fission process are neutrons and fission products. Between two and three neutrons on the average are created for each fission event. It is these neutrons which enable the fission chain reaction to be self-sustaining.

months in the fuel storage pool at the plant site for cooling. Following cooling and pending the availability of reprocessing capacity, the fuel is either stored on the plant site or moved to an interim storage facility. This service is estimated to cost $15/kg U per year. A storage facility capable of supporting 10 power plants is estimated to take four years to construct and cost $45 million. Therefore, the investment per reactor supported is $4.5 million or $4.5/kw.

Spent Fuel Shipment

The movement of discharged fuel from the power plant to an interim storage facility or reprocessing facility requires a shipping container that is capable of providing shielding for the radioactive materials now contained within the fuel and cooling for the residual heat produced by the radioactive materials.

The container is required to be able to maintain its integrity under severe accident conditions including a 30 foot drop, immersion in water, and fire. These requirements result in shipping containers weighing, when loaded, upwards of 100 tons. The current costs of spent fuel shipping is approximately $25/kg.

Reprocessing of Spent Fuel

Following storage for cooling, the spent fuel is shipped to a reprocessing facility to reclaim the residual uranium and plutonium and to concentrate the radioactive fission products. The process is generally in three steps:

1. Head end treatment
2. Separation or extraction
3. Product purification

The head end treatment prepares different types of fuel elements for reprocessing. The practice is to remove the cladding from the fuel either mechanically or chemically and then to dissolve the fuel material in nitric acid.

A number of different methods can be utilized to extract uranium and plutonium from the fuel solutions.* The uranium and plutonium thus recovered can be used as feed material for additional fuel. Radioactive waste from these processes must be solidified in preparation for their ultimate disposal. Because of current uncertainties regarding licensing requirements, it is difficult to accurately assess reprocessing costs for a new facility. Depending on the financing structure for such a venture, cost estimates of from $150 to 400/kw have been made.

A facility capable of supporting 50 power plants is estimated to take ten years to design and construct and cost $700 million. The investment per plant is, therefore, $14 million or $14/kw.

---

*These methods include both solvent extraction and continuous ion exchange. The product of the extraction step is a nitrate solution of plutonium and uranium. The uranium is recovered either as a nitrate (liquid) or as UF_6 (solid). The plutonium is recovered either as a nitrate (liquid) or as oxide (solid).

Waste Disposal

Upon discharge from the reprocessing facility, radioactive wastes are solidified into an insoluble, non-leachable form. The solid waste thus produced is then packaged and ultimately stored. It is well known that there are questions surrounding the ultimate storage of radioactive wastes. The problem results from the fact that the waste is highly radioactive for many thousands of years; therefore, the time during which the wastes must be controlled and separated from the biosphere is long relative to the duration of all human created systems (e.g., governments).

The long-term approach has been to attempt to isolate wastes in geologic formations, whose stability has been demonstrated over hundreds of thousands of years. The most promising of these formations are bedded salts. After a number of false starts, ERDA is currently designing a pilot facility for the storage of wastes in bedded salt formations. Upon satisfactory completion of the pilot plant, it is then planned to construct a commercial sized facility. The current concepts include the ability to retrieve the wastes should new approaches to waste storage be developed following the initial placement of the wastes.

It is estimated by ERDA that a facility capable of supporting 200 power plants would cost \$56 million or \$2.8 million per plant. Current estimates of the cost of this service are \$50/kg.

Uranium Recycle

The uranium and plutonium recovered in the reprocessing facility may be utilized as feed material for additional nuclear fuel.* Feed requirements are reduced by uranium recycle by 1/6 or 17%. There are no significant SWU savings through uranium recycle.

---

*As indicated above, discharged fuel consists of approximately 7 gms U^{235} and 5 gms Pu per kilogram of discharged uranium. The fuel and enrichment requirements for 1 kilogram of 2.8% enriched uranium at 0.3 w/o tails are found below.

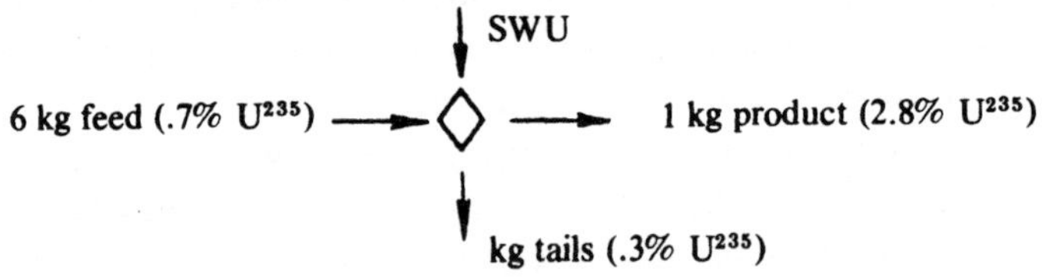

Feed and SWU requirements are reduced in the recycle mode in the following manner:

Uranium recycle:

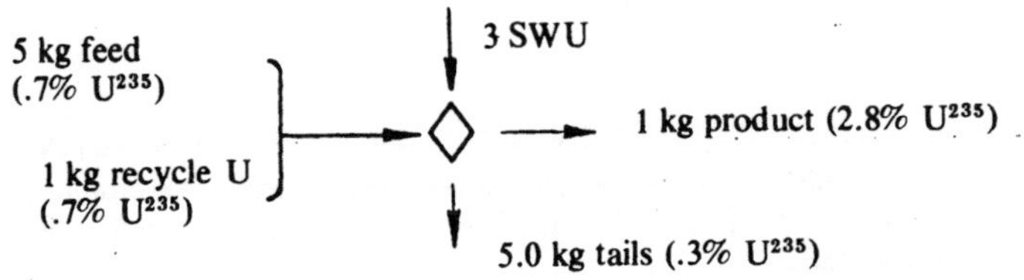

Plutonium recycle

To evaluate plutonium recycle benefits, one assumes 1 gm Pu is equivalent to .85 gms U^{235}.* This enriched uranium equivalent then reduces the need for enriched product by approximately 20%.

Feed is therefore reduced by 1/6, as in the uranium recycle case, *and* SWU requirements are reduced by 1/5 or 20%.** *Thus the savings inherent in the recycle of uranium and plutonium are summarized as follows:*

Uranium feed	33%
SWU	20%

Uranium feed savings result equally from U and Pu recycle, while SWU savings exclusively result from Pu recycle. While the savings in SWU and uranium through recycle are easily calculated, the economic incentive to do so is not easily analyzed. The economic decision is evaluated through the following equation:

Uranium value	+	plutonium value	⩾	Reprocessing cost	+	waste disposal cost	−	fuel storage cost

If one assumes that the fuel storage cost (i.e., in the event reprocessing is not performed) is equivalent to the waste disposal cost (this is not unreasonable in light of recent studies), then the economic judgment reduces to:

$$\text{Uranium value} + \text{plutonium value} \geqslant \text{Reprocessing cost}$$

Uranium value is dependent upon the cost of ore, conversion and enrichment; plutonium value includes consideration of these factors as well as cost penalties associated with safeguards and additional fabrication costs.

The relationship between "Breakeven Reprocessing Cost" (BRC) and

---

* Therefore, if discharge fuel contains 5 gms Pu, it is equivalent to 4.25 gms U^{235}. The U^{235} equivalent is then blended with natural uranium to make up 2.8% enriched material in the following manner:

4.25 gms U^{235} + .200 kg natural U (1.4 gms U^{235})
= .204 kg U containing 5.6 gms of U^{235} equivalent
= .204 kg 2.8% enriched U

**Our feed/SWU diagram for the Pu recycle case then becomes:

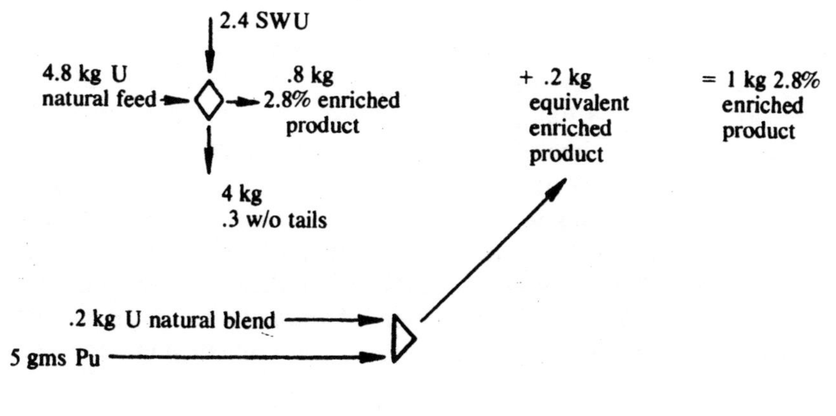

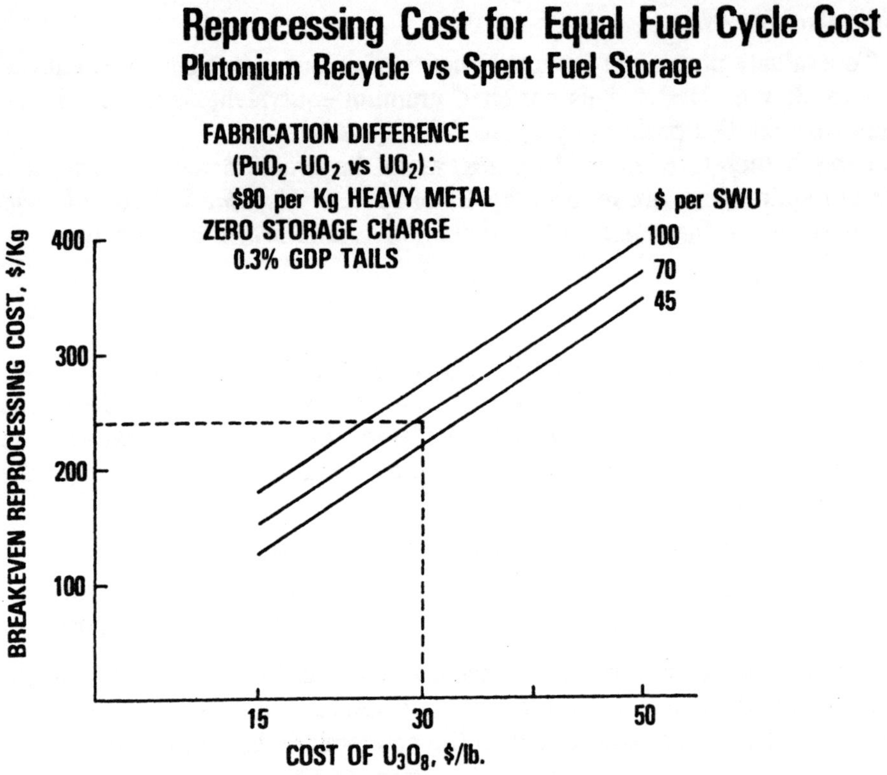

uranium and SWU cost for a given plutonium fabrication penalty is shown (above), taken from ERDA-33. If reprocessing costs are greater than BRC, then reprocessing and recycle are "uneconomic". If reprocessing costs are less than BRC, then reprocessing and recycle are "economic".

It should be emphasized that there may be reasons *for* reprocessing such as mandatory waste concentration and reasons for *not* recycling plutonium such as *safeguards* which are not considered in the economic evaluation presented above.

Nuclear Fuel Cycle Investment

The fuel cycle investment required to support a 1000 MWe nuclear power plant is summarized below by fuel cycle category.

Considering variable uranium requirements, ore grades and other factors, the total full fuel cycle investment required to support one 1000 MWe plant is in the order of $70 to 100 million. The cost of the nuclear power plant is estimated to be approximately $1 billion, and therefore, fuel cycle investments represent 7 to 10% of the total required nuclear industry investment. The argument has been made that since this is a relatively small fraction, the fuel cycle investment will be there if the investment in the power production facilities is there. This might be true if the risk of these investments were the same. It is noted that the development of uranium reserves and mining capability is a relatively high risk venture. Enrichment and reprocessing,

Table B.1.

Fuel Cycle Component	Facility Capital Investment × 10^6	Number of 1000 MWe Reactors Supported	Investment per Reactor × 10^6
Exploration, Mining & Milling (100 tons U_3O_8/year)	66 (a)	6	11 (a)
Conversion (5000 tons U_3O_8/year)	24-79 (b)	30	.9-1.3 (b)
Enrichment (9,000,000 SWU/year)	3,360-4,500 (b)	120	28-37.5 (b)
Fuel Fabrication (1,500,00 kg/year)	90-155 (b)	50	1.8-3.1 (b)
Fuel Storage	45	10	4.5
Reprocessing (1,500,000 kg/year)	700	50	14
Waste storage	560	200	2.8

(a) Depending on ore grade and whether or not there is recycle.
(b) Depending upon whether or not there is recycle.

because of their capital intensive nature and large facility size requirements, involve large risks associated with facility downtime. Because of the long lead times associated with these investments, questions as to the timely availability of materials and equipment become a problem. It is invalid then to assume that merely because fuel cycle investments are a small piece of the total financing problem, that they will be assured.

Nuclear Fuel Cycle Economics

The computation of nuclear fuel cycle costs is performed in a manner similar to the computation of fossil fuel costs. In the case of fossil fuels, the cost of the material, in the case of coal, e.g., \$25/ton is simply divided by its energy content, e.g., 12,500 BTU/lb. The result of this computation \$25/ton/(12,500 BTU/lb) (2,000/ton) = \$1.00/MBTU is the fuel cost of the material.

The development of material costs for nuclear fuel is performed as follows:

Uranium Feed

Assume yellowcake costs \$25/lb, conversion costs \$1.25/lb

\$25/lb yellowcake (U_3O_8) × 1.18 lb U_3O_8	= \$29.50/lb U
Conversion to UF_6	= \$ 1.25/lb U
Total	= \$30.75/lb U
	= \$67.75/kg U

Note that the cost developed thus far is for k kilogram of natural uranium feed. The section above on enrichment indicates that for enrichment plant operation at 0.3% tails, 6 kilograms of natural uranium feed are required for 1 kilogram of 2.8% enriched product. Therefore, the cost of feed is 6 × $67.65 or $406/kg U.

Enrichment

It is noted above that 3 SWU are required per kilogram of 2.8% enriched uranium. At a SWU cost of $75/SWU, the total cost of enrichment is $225/kg U.

Fabrication

The cost of fabrication is assumed to be $80/kg U.

Interest Costs

Since the investment in nuclear fuel is maintained over a period of years, it is appropriate to consider the cost of that investment as a component of fuel cycle costs. The total initial investment in fuel is:

Feed	$406.00
Enrichment	225.00
Fabrication	80.00
	$711.00/kg U

If this investment is amortized over four years at a rate of 10%/year, the interest component of fuel cycle cost is computed to be $142/kilogram.

Spent Fuel Storage

Spent fuel storage costs are assumed to be $25 kg.

Spent Fuel Shipping

Spent fuel shipping costs are assumed to be $25/kg.

Reprocessing

Reprocessing costs are assumed to be $200/kg U.

Waste Disposal

Waste disposal costs are assumed to be $50/kg U.

Summary

The fuel cycle cost computation involves taking the total cost of the fuel and dividing by its energy generation. Because of the uncertainty surrounding the tail end of the fuel cycle, costs of reprocessing and uranium and plutonium credits will not be included. It is assumed in the calculation that the cost of permanent spent fuel storage is equivalent to the cost of waste disposal.

Feed	$406
Enrichment	225
Fabrication	80
Interest	142
Interim Storage	25
Transportation	25
Waste Disposal	50
Total	$953/kg U

A kilogram of uranium can be expected to produce 600,000 kw hrs or 2000 MBTU over its four year residence.

The total fuel cycle in this case is computed to be:

$$\frac{\$953/\text{kg U}}{2{,}000\ \text{MBTU/kg U}} = \$0.48/\text{MBTU}$$

III. Why the "generation mix" for electricity production?

The selection of the type of capacity to be installed in any utility system for the generation of electricity depends on the system load requirements (how long it operates at what level of capacity) and the projected economics of the type of electricity generating system being considered. The types and amounts of capacity selected for a given generating system are referred to as the "generation mix".

The demand for electricity is not uniform. A typical system load duration curve is shown in *Figure B.2*. This load duration curve indicates that demand falls into three sections: peak, intermediate, and base. Peak loads are for short duration in the course of the year and require low fixed cost equipment to be available for service; in contrast, base loads are for many hours during the year, requiring units with low variable costs. Figure B.2 shows the duration of each type of load as a function of the number of hours per year it persists.

The selection of capacity to serve such a load curve is determined by the economic characteristics of various forms of generation. A load screening curve is utilized for such analysis. A screening curve shows how the annual dollar cost of a kilowatt hour depends upon the number of hours that each form is used for generation. If no electricity at all is generated, the capacity in place has an annual cost which is carrying the cost of generation plant investment. Power generation then adds the variable costs of fuel, load factor and maintenance. *Figure B.3* shows screening costs for nuclear, coal, oil-combustion turbines, and pumping storage plants.

Costs utilized are shown below and correspond to actual projected 1977

Figure B.2.

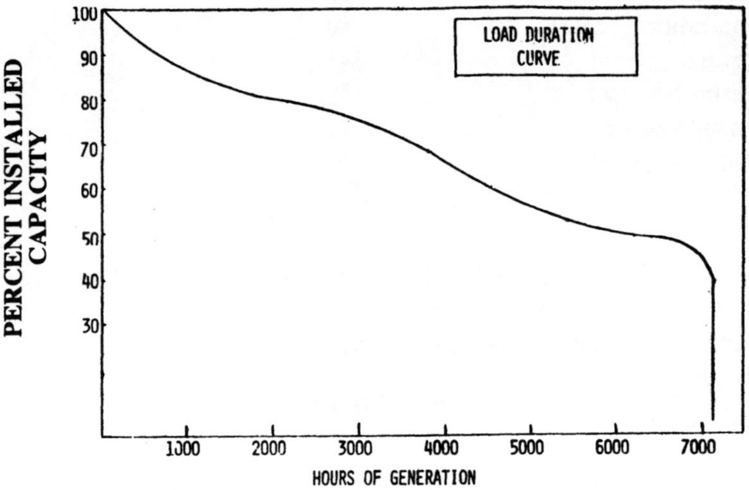

Figure B.3.

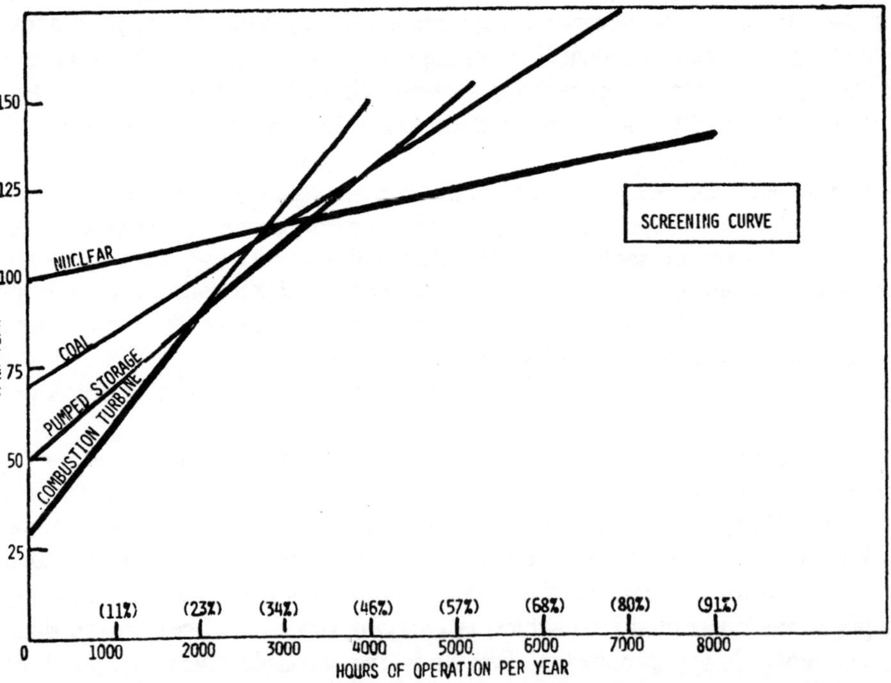

costs of an existing 800 MWe nuclear unit, and a nearby 650 MWe coal fired unit, with fuel costs estimated for 1980. Combustion turbine costs and pumped storage costs are extrapolated from current experience. Oil costs are 1980 estimates, while pumped storage fuel costs are assumed to be 1.33 times the cost of coal.

Plant (capital cost)	Fixed costs* $/kw-year (15% fixed charge rate)	Variable costs** $/kw-year/1000 hrs of generation
Nuclear ($666/KWe)	100	5
Coal ($467/KWe)	70	15
Combustion turbine ($200/KWe)	30	30
Pumped storage ($333/KWe)	50	20

* Included are: costs of plant only
**Included are: costs of fuel only; i.e., does not include costs of operation and maintenance.

It is seen from Figure B.3 that from 0 to 2000 hours of operation, economics favor gas turbine capacity; from 2000 to 3400 hours, pumped storage is favored; and from 3400 and beyond, nuclear is favored. In reality, some capacity diversity should be developed, so some coal generation could be installed to complement the nuclear base load. To determine the amount of each type of capacity which should be installed, one simply goes back to the load duration curve.

IV. Postscript on Plutonium

Plutonium is a transuranium element not found in nature except in trace quantities. It was discovered in 1941 by G. T. Seaborg and his associates. There are fourteen isotopes of plutonium with mass numbers between 232 and 246; the isotope of greatest interest is plutonium 239. Plutonium 239 is formed in nuclear reactors in the following manner:

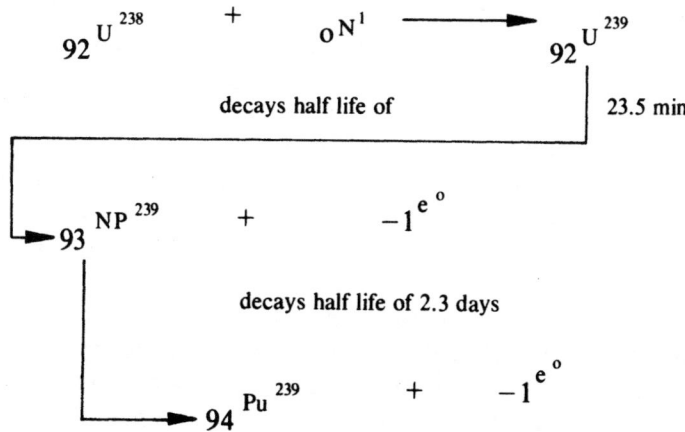

$$_{92}U^{238} + {}_{0}N^{1} \longrightarrow {}_{92}U^{239}$$

decays half life of 23.5 min

$$\longrightarrow {}_{93}NP^{239} + {}_{-1}e^{0}$$

decays half life of 2.3 days

$$\longrightarrow {}_{94}Pu^{239} + {}_{-1}e^{0}$$

The common shorthand representation of the reaction being:

$$U^{238} + n \longrightarrow Pu^{239}$$

Pu^{239} is an alpha emitter with a half life of 24,400 years. It is also readily fissile and can support a fission chain reaction in the manner of U^{235}.

The fissile nature of plutonium makes it both a benefit and a problem in light water reactors. As plutonium is being produced in the reactor fuel, it can be utilized as a fissile material. Thus, in normal operation, 20 to 30% of the fission events occurring are in plutonium. The presence of plutonium enhances the economics of the fuel cycle even without recycling.

The recycling of plutonium can result in uranium resource savings of approximately 16% and SWU savings of approximately 20%. Thus, if reprocessing costs warrant it, Pu recycle can reduce fuel cycle costs.

The fissile characteristics of plutonium coupled with the ability to chemically separate it from uranium result in the potential for plutonium use as a weapons material. Plutonium discharged from light water reactors once separated from uranium is a relatively high grade fissile material. This product, while not as high grade as "weapons grade plutonium," can be used in nuclear weapons. The separation process (reprocessing) is difficult and complex in an engineering sense because of the toxicity of the plutonium and fission products contained in the discharged fuel, but the basic chemistry is well established. Comparable separation and concentration of the fissile uranium isotope, U^{235} can only be accomplished through enrichment. Enrichment technology is classified in most countries, and enrichment plants are large, complex, and expensive. Until the recent development of centrifuge technology, large amounts of electricity were required.

The radioactive character of plutonium results in an additional toxicity problem associated with its utilization. The radiobiological hazards of plutonium were recognized quite early and concern for them has resulted in extensive research on internal deposition of plutonium and its effects on animals and man. The organs of primary interest are the lungs, lymph nodes, liver, and bones where radiation subsequent to deposition of plutonium can cause biological damage. Inhalation followed by translocation is probably the most important path to these tissues. Controversy exists over what the precise limits of plutonium exposure should be. "Hot particle theory" advocates recommend substantial reductions in existing plutonium exposure limitations. Plutonium, while quite hazardous to handle, is not "the most toxic material on earth." It is less toxic for example than such materials as shellfish toxin or botulism.

Such arguments, however, have only little relevance to the real technical issues surrounding the utilization of plutonium. In order to safely utilize this potentially beneficial, potentially dangerous element, adequate safeguards for separating plutonium from the biosphere and keeping it out of the hands of those who would use it for weapons are required. The solution of these problems is not technical. The technical means for accomplishing the above objectives already exist. The problem is a political/societal one related to adequately assuring the people that the potential risks associated with plutonium are well covered and that any remaining risk (if there is a remaining risk) is worth the potential benefit from the utilization of plutonium.

APPENDIX C

A SUMMARY OF NUCLEAR FUEL TRADE POLICIES AND PRACTICES

The significance of trade among nations for nuclear fuel is illustrated by the fact that today only nine countries have major deposits of uranium ore and currently only four nations or groups of nations seek to provide other nations uranium enrichment services as contrasted to the many nations with nuclear power plants in operation, under construction or planned.

The trade policies and practices of the nations in a position to export uranium or low enriched uranium at the beginning of 1976 are discussed in this section of the study. This review outlines the requirements and restrictions imposed by these nations as of that time. It should be emphasized that changing conditions of supply, costs, world concern and other factors can result in major changes to established government policies.

I. Exports of Uranium

The nations of the world with significant deposits of uranium ore include the following, and their trade policies and practices will be discussed in turn:

Australia	South Africa
Canada	Sweden
France	United States
Gabon	U.S.S.R.
Niger	

1. Australia

The Australian Government has approved exports contracts to sell 8,863 metric tons (tonnes) of uranium (about 11,522 short tons of yellowcake) for delivery between 1974-86.[1] Recent indications confirm that Australia is now prepared to resume contracting for the export of large quantities of uranium beginning in 1979. By 1980-81, Australia could have production capacity for up to 28,000 tons of uranium annually, though actual production for export is projected at about half that amount. (Some Australian mining industry representatives contend that this level of production could be reached as early as 1979.)

Although existing contracts for export were written in terms of U_3O_8, Australia as of mid-1976 was in the process of considering the feasibility of a UF_6 plant, and within another decade may decide to undertake to export

[1] Twenty-Second Annual Report, Australian Atomic Energy Commission (June 30, 1974)

some enriched uranium. New contracts for the delivery of Australian uranium will probably contain provisions for the possibility of eventual delivery of some proportion of the uranium as enriched product.

The Australian requirements for the export of uranium reflect the Australian Government's concern and commitment to non-proliferation as well as commercial and energy policy interest. The Australian Government's authority over uranium exports is based on export control and the Atomic Energy Act. Australia is not a member of the Nuclear Exporters Group, but it can be expected to require safeguards broadly consistent with those of other exporters. While all exports must be accompanied by safeguards, the precise terms are under review.

On April 7, 1976, the Australian Government made the following progress report on that review:

> "The export of Australian nuclear material under any future contracts will need to be consistent with Australia's obligations under the Treaty on the non-proliferation of nuclear weapons (NPT) and associated arrangements to which Australia is a party. These arrangements require the application of safeguards administered by the International Atomic Energy Agency (IAEA) to material supplied by Australia to all non-nuclear-weapon states. They also provide for the application of IAEA safeguards to Australian material should it be re-exported to such countries from nuclear-weapon states.
>
> "The Government intends, in addition, that the Government of any country wishing to import Australian uranium will be asked to conclude with the Australian Government a prior bilateral agreement. This agreement will then provide direct, formal assurances from the importing country to the Australian Government that material supplied by Australia will not be diverted from the purposes specified in the agreement. It will also specify appropriate safeguards and controls to this end.
>
> "In considering the details of the safeguards and controls which importing countries will be asked to accept, the Government has under consideration a range of proposals, to which it is giving full and careful attention. At this stage it has in mind that it will, in negotiating future bilateral agreements, ask that the importing country accept the following provisions in those agreements for the export of Australian nuclear material:
>
> - Provisions fully reflecting Australia's obligations to ensure the application of safeguards to verify that material supplied to any non-nuclear-weapon state for peaceful purposes is not diverted to any explosive use,
> - provision for the continued application of appropriate IAEA safeguards in non-nuclear-weapon states party to the NPT in

the event that safeguards under the NPT should for any reason cease to apply at any time in the future,

- provision for Australia to reserve the right to apply other safeguards in the event that IAEA safeguards should for any reason cease to apply at any time in the future in non-nuclear-weapon states,
- in general, agreement by the importing country that material supplied by Australia will not be consigned to another country (including consignment for upgrading) or re-exported without the Australian Government's consent which could be given at the time of sale.

"The Government expects that a clause will be included in all future commercial contracts for the export of Australian nuclear material making it clear that transactions are subject to safeguards as agreed between Australia and the importing country.

"The Government wished there to be no doubt that Australia will continue to support the strengthening of international restraints on nuclear weapons proliferation. In line with the recommendations adopted by consensus by all countries which attended last year's NPT Review Conference, we shall support constructive multilateral efforts to encourage the application of IAEA safeguards on all peaceful nuclear activities in importing countries not party to the NPT. Australia's safeguards policy in relation to exports of nuclear materials will be kept under review, to take into account changing circumstances and on-going international efforts to strengthen controls against the proliferation of nuclear weapons."

2. Canada

The Canadian Government on September 5, 1974 announced a new uranium policy with two primary objectives:

(a) To insure at least a 30-year reserve of nuclear fuel for all existing, committed and planned reactors in Canada for any ten-year forward period.

(b) To ensure that sufficient uranium production capacity is available for the Canadian domestic nuclear power program to reach its full potential.[2]

The total domestic in-service nuclear capacity by 1985 is estimated to be approximately 18,400 MWe requiring 92,000 tons of U_3O_8 through 2002. This amount will be increased annually as new domestic reactors are planned. Contractual export commitments exist for about 120,000 short tons of U_3O_8. Most of this amount has been approved for export subject to the signing of necessary safeguards agreements.

It is estimated that over and above the export commitments and domestic

---

[2]Energy, Mines and Resources Canada, "1974 Assessment of Canada's Uranium Supply and Demand," p. 3 (August 1975).

allocations, almost 50% of the Canadian uranium resources are uncommited for future export or domestic needs. The present expectation is that between now and 1985 there can be firm export commitments, between 1985 and 1990 conditional export commitments, and after 1990 no exports because of domestic requirements.

Canadian exports must be accompanied by IAEA safeguards, an export license and a statement of intended end use. Added physical security measures are expected to be required.

3. France, Gabon and Niger

The French Government manages the exports of uranium for Gabon and Niger so the three can be considered together. However, Gabon and Niger do have some direct contracts with Italy, Germany and Japan. The French have stated as an objective supplying 20% of the world's nuclear fuel needs with a great majority of the fuel coming from Gabon and Niger. Additional information concerning the French policy on uranium trading has been requested from URANEX.

4. South Africa

The Atomic Energy Commission of South Africa has stated its position as follows:

> "In common with other uranium producing countries, the South African Government exercises, through the provisions of the Atomic Energy Act, No. 90 of 1967, control over the disposal and export of uranium in whatever form. Section 7(1) of that Act provides that, except with the written authority of the South African Minister of Mines, no person shall dispose of or export uranium from the Republic of South Africa. Any authority under section 7(1) may be given subject to such conditions as the Minister may in his discretion impose. Such conditions relate to quantity, price, the form in which the uranium can be exported and safeguards arrangements. When dealing with these matters it is customary for the Mininster to act on the recommendation of the South African Atomic Energy Board.
>
> "Sales of uranium by South African producers to non-nuclear weapon states are made subject to the safeguard provisions of the International Atomic Energy Agency or to equivalent safeguards arrangements which are acceptable to my Board in order to ensure that the nuclear material will be used for peaceful purposes only.
>
> "Where the sales agreement has been approved by the Minister, producers are required to obtain specific export authorities (licenses) covering the export of individual or annual consignments. An export authority will only be issued if the conditions and requirements imposed by the Minister have been fully met and the Board has been satisfied on the end-use of the material.[3]"

[3]Letter from Mr. A. J. A. Roux dated February 11, 1976.

5. Sweden

Sweden at the beginning of 1976 was not exporting uranium and is not expected to do so in the near future.

6. United States

The United States has an "open policy" with regard to the export of unenriched uranium. Thus, no request for the export of yellowcake or UF_6 has been turned down.

As of January 1, 1975 the following foreign uranium commitments by domestic producers had been made:

Year of Delivery	Tons U_3O_8
1966 thru 73	5,500
74	1,500
75	600
76	500
77	1,400
78 and 79	1,100
Total	10,600[4]

The requirements include an Agreement for Cooperation, IAEA safeguards and an export license. The question of physical security measures is still an open one.

With regard to imports to the United States of foreign uranium for domestic use, plans have been announced to begin the gradual removal of the enrichment restrictions in 1977. At that time 10% of the feed furnished by a customer under all of its enrichment contracts may be of foreign origin, with increasing percentages until restrictions are removed in 1984. As of January 1, 1975, the uranium delivery commitments for import by domestic buyers were as follows:

Year of U_3O_8 Delivery	Annual Tons	Cumulative Tons
1975	800	800
1976	1,500	2,300
1977	2,600	4,900
1978	3,100	8,000
1979	3,000	11,000
1980	2,700	13,700
1981	3,500	17,200
1982	3,700	20,900
1983	3,600	24,500
1984	3,600	28,100
1985	3,400	31,500
1986	2,300	33,800
1987-1990	2,000/yr.	41,000[5]

[4]Energy Research and Development Administration No. 24, "Survey of United States Uranium Marketing Activity," p. 6, (April 1975).

[5]*Id.* at p. 8.

7. U.S.S.R.

Russia does not export any yellowcake or UF_6. Its policy is to export only enriched uranium.

II. Exports of Enriched Uranium

The sources at the beginning of 1976 for the international trading of enriched uranium in the world include the following:

ERDA (United States)[6]
Eurodif (France, Spain, Italy, Belgium and Iran)
URENCO (West Germany, Netherlands and United Kingdom)
U.S.S.R.

ERDA (United States)

The current policy concerning the export of low enriched uranium by ERDA requires (1) an Agreement for Cooperation between the United States and the importing nation, (2) the establishment of IAEA safeguards or bilateral safeguards[7] with respect to the exported enriched fuel, (3) a physical security review by ERDA of the importing nations' fuel protection systems, (4) a statement by the importing country concerning intended use of the enriched uranium, and (5) an export license issued by the Nuclear Regulatory Commission.

As of 1976 the vast majority of the enriched uranium being traded throughout the world was supplied by ERDA under contracts negotiated by ERDA. These contracts have the following non-commercial provisions:

1. The contract provides that the agreement "shall be in all respects subject to and in accordance with all the terms and conditions of the Agreement for Cooperation, as it may be amended, it being understood that the guarantees and safeguards provided for therein will always apply to any material transferred hereunder except to any material after subsequent transfer by the Customer, with the approval of the Commission, to a nation or group of nations with which the Government of the United States of America has an Agreement for Cooperation within the scope of which such subsequent transfer falls." (Item 14 of "General Terms and Conditions.")

2. The Contract provides that the customer "shall procure all necessary permits or licenses (including any special nuclear material licenses) and comply with all applicable laws, treaties, regulations, and ordinances." (Item 11 of "general Terms and Conditions.")

3. The contract provides that there shall be no restriction on the customer's use or disposition of the furnished enriched uranium, provided that the uranium may not be retransferred beyond the jurisdiction of the recipient

[6]The Ford Administration has proposed legislation to Congress for the purpose of encouraging the development of private enrichment in the United States which, if adopted and used by private industry, would add new sources for providing enriched uranium.

[7]At the moment bilateral arrangements apply only to Italy until 1978, after which all will be subject to IAEA.

country except as authorized in accordance with the Agreement for Cooperation. (Art. II.2.d.)

4. The contract provides that the US can terminate the agreement if delivery of enriched uranium would be inconsistent with obligations assumed by the US under the NPT. The customer has the same right in the event that deliveries of feed material would be inconsistent with its obligations under the NPT. (Art. IX.5.)

Eurodif (France, Spain, Italy, Belgium and Iran)

Eurodif as of January 1976 was not exporting any enriched uranium but was in the process of contacting nations for exports beginning in 1978-1979. The Eurodif proposals require IAEA safeguard provisions and a statement as to the intended use of uranium. In connection with physical security measures, it is believed that Eurodif requires that due consideration be given to physical security. The Eurodif contract itself does not contain a provision regarding safeguards nor one regarding permits and authorizations.

URENCO (West Germany, Netherlands and United Kingdom)

URENCO as of January 1976 was not exporting enriched uranium but it was in the process of writing contracts for this purpose. Its requirements appear to be identical to those of Eurodif in that they require IAEA safeguards, a statement concerning the intended use of the uranium and due consideration to physical security measures. The URENCO contract contains several non-commercial provisions:

1. The contract provides that product material delivered under the agreement "shall be subject to safeguarding procedures, regulations and control in accordance with the provisions of Chapter VII of the Treaty establishing the European Atomic Energy Community if appropriate, or otherwise in accordance with all relevant obligations and undertakings of the [two] Governments . . . . Product material delivered to the Customer in accordance with the provisions of this Agreement shall be used for peaceful purposes only." (Clause 12.)

2. The contract provides that the supplier shall be responsible for obtaining all licenses, permits and other authorizations regarding the construction and operation of the enrichment plant. The customer shall be responsible for obtaining all licenses, permits and other authorizations regarding the construction and operation of the power plant as well as obtaining all import and export licenses. It should be noted that the URENCO contract further provides: "Failure to fulfill any of the obligations set out in this Clause, whether through the refusal of any authority to issue a license or through any other cause, shall not be deemed to be a circumstance outside the control of either party for the purposes of [the force majeure clause]. (Clause 13.)

USSR

The USSR has contracts with West Germany, France, Italy and Spain to provide enriched uranium. They are in the process of contacting nations such as Japan for the purpose of additional sales. It is understood that they require IAEA safeguards and a statement as to the intended use of the uranium. Physical security measures are not required. The USSR contract does not contain a provision regarding safeguards but it does provide that the "receipt of the import licenses is a duty of the Consignee." (Art. 12.1)

Possible Additional Countries

Several additional sources of enriched uranium can be expected to develop in years ahead. The announced West German-Brazil Agreement provides the potential for Brazil to enter into the world market for the purpose of exporting enriched uranium. South Africa is expected to become an exporter of enriched uranium before too long in light of UCOR and the recent Iran-South Africa agreement. Australia may, within the next decade, decide to export enriched uranium. Canada has given consideration to the possibility of establishing an enrichment capability for export purposes.

APPENDIX D

GLOSSARY OF ACRONYMS

Acronym	*Full Name*
BTU	British thermal unit
BWR	Boiling water reactor
CANDU	(A type of heavy water reactor)
EBR	Experimental breeder reactor
EEC	European Economic Community
EEI	Edison Electric Institute
Eurodif	(Uranium enrichment consortium lead by France, and including Italy, Spain, Belgium, and—indirectly—Iran)
FFTF	Fast flux test facility
GA	(Gulf-Shell) General Atomics
GCBR	Gas cooled breeder reactor
GCR	Gas cooled reactor
HTGCR	High temperature gas cooled reactor
HTGR	High temperature gas reactor
IAEA	International Atomic Energy Agency
IBRD	International Bank for Reconstruction and Development
LMFBR	Liquid metal fast breeder reactor
LWBR	Light water breeder reactor
LWR	Light water reactor
MTHM	Metric tons heavy metal
MWe	Megawatts of electricity
NERC	National Electricity Reliability Council
NPT	Non-Proliferation Treaty
NURE	National uranium resource evaluation
OECD	Organization for Economic Cooperation and Development
OPEC	Organization of Petroleum Exporting Countries
OPEN	Organization des Producteurs d'Energie Nucléaire
PWK	Plannungsgesellschaft für Wiederaufarbeitung von Kernbrennstoff
PWR	Pressurized water reactor
SALT	Strategic Arms Limitation Treaty
SIPRI	Stockholm International Peace Research Institute
STU	Short ton unit
SWU	Separative work unit
Unirep	United Reprocessors Limited
URENCO	(Centrifuge partnership: UK, Germany, Netherlands)
US ERDA	United States Energy Research and Development Administration

APPENDIX E

COMMENTS AND DISSENTS OF THE WORKING GROUP

STATEMENT, by David E. Lilienthal

While I do not dissent or wish to comment adversely on the conclusions and recommendations in this report, I do believe that there should be a stronger sense of urgency in seeking solutions to the proliferation problem. I am concerned that the current consensus while identifying the problem has generally adopted the view that nothing much can be done to solve it. In my view, solution is critical to the future of mankind and deserves concentrated attention, nationally and internationally, at the earliest possible time.

I would like to reiterate observations I offered on this point at a recent meeting of planning officials in Washington.

"Our country in 1945 had an atomic monopoly. American presented a plan to avoid an atomic arms race and prevent the spread of nuclear weapons by internationalizing nuclear technology and development. The objective was of the greatest possible importance, both to ourselves and to the rest of the world. Everyone recognized that atomic arms proliferation might be catastrophic; the American plan of 1946 was intended to see that this did not happen. No one, including the Soviet Union, did or could question the objectives of that plan, namely to free the world, both the Free and the Communist world, of the ugliest cloud ever to hang over all mankind.

"But when the American plan failed to find acceptance by the Soviet Union, for whatever reason, we then turned, in time, to an alternative plan, with the bitterly ironic name of "Atoms for Peace"! What have been the results of this alternative plan, diametrically opposed to the objectives of America's original plan? The atomic arms race has become more furious and more insane than ever. Atomic warfare is today actually closer than ever. And in the past few weeks we are presented with this spectacle: French commercial interests, using American-derived technology and materials, have agreed to provide the nervous, uneasy Pakistanis with nuclear capabilities that can enable them to make atomic weapons. The Pakistanis, meanwhile, are reminding us pointedly that India already has atomic weapons capabilities, and that the Indian capabilities originated in the materials from nuclear reactors which we persuaded the Indians to buy from us years ago, under the alternative Atoms for Peace plan.

"How tragically short are such baleful results as these of the high intentions of those who promoted the alternative plan!

"The great hazard at the present time is that the sense of resignation in the seats of American authority, and among the American intellectual elite, and the complacent views as expressed in testimony recently before the Senate

Committee on Governmental Operations will become pervasive and the world will accept their view, as expressed recently, that there is now no choice but to continue to follow the present disastrous course. With this I profoundly disagree. The stakes are too high, the urgency too great, for such complacency or weak-kneed resignation, such bankruptcy of ideas. We the people of the world do have a choice."

David E. Lilienthal
Chairman, Development and Resources Corporation; Former Chairman, Atomic Energy Commission.

COMMENT AND DISSENT, by Laurence I. Moss

The report would have been more useful, and less misleading, if it had given greater recognition to the importance of price in establishing future demand for electricity and hence the "need" for nuclear and/or coal powerplants. The demand projections used in the report are considerably higher than those estimated in recent competent, independent studies not cited therein. Furthermore, if a policy decision were made to price electricity on the basis of *incremental* rather than *average* cost, so that a meaningful comparison of marginal benefits and costs could be made by users, the growth rate would be lower still.

The point bears repeating: the present basis of price regulation results in user decisions based on perhaps $200 to 300 per installed kilowatt capital investment for generation, whereas the utility expands its generating capacity at cost of $800 to 1100 per kilowatt. Thus, the nation's scarce resources are squandered. There is too much investment in new generating capacity, and not enough investment in more efficient industrial processes, more insulation of buildings, more efficient systems for heating and cooling, more use of solar energy in low-temperature heating applications, etc.

This observation leads to two recommendations which should have been made. The first is to urge that steps be taken to implement a policy of incremental cost pricing of electricity (as well as gas and oil), with revenue in excess of that normally allowable by the regulatory commissions rebated in a manner designed to not encourage uneconomic uses of energy. The second recommendation is that production costs should not in any way be subsidized by the government, through ownership, guarantees, or other devices, because these serve to shift part of the full cost of energy from where it belongs (on the user, in relation to his use) to where it clearly should not be (on the taxpayer, in relation to his tax burden). Note that this recommendation does not argue against government subsidy of research and development, because a well-known market failure would result in less than the socially optimum investment in R & D if it were left to the private sector.

With these policies in effect, the growth rate in demand for electricity (and other forms of energy) would be sufficiently moderate to permit us the luxuries of both time and choice. That is, we could select among the various energy supply options those which were least harmful to health and the environment, rather than being forced to pursue both the good and bad at great haste. And the economy would benefit because our limited resources would be deployed in a manner yielding, on average, a higher rate of return.

This comment is not meant to imply that, given the choice, nuclear should be abandoned in favor of coal. The available evidence simply does not support such a conclusion. To the contrary, it is likely that the environmental, health and social costs of the use of coal are greater than the comparable costs for nuclear, even assuming that human error, mechanical failure, sabotage, and terrorism will sometimes occur.

With respect to certain other technologies, the report is misleading in deprecating these technologies on the basis of not being capable of providing large supplies of energy in the near or intermediate term (now to about 1990). But much of what is recommended to encourage nuclear would not have a major impact in that same period. The point is that both the near and long term should be considered in evaluating alternatives and developing recommendations; this was done for nuclear, but not for solar and geothermal.

One recommendation included in a draft of the report but omitted in the final version was the following: "Nuclear fuel exports should be returned for reprocessing either to the country of origin or to a third country mutually agreed upon by the nuclear fuel exporter and importer." I believe this recommendation should have survived. The diversion of nuclear fuel suitable for weapons purposes by the government of non-nuclear nations is one of the most troublesome risks of nuclear power. The above recommendation, if implemented, would reduce that risk. It would not, to my knowledge, be inconsistent with related actions taken by supplier nations in informal (and to date secret) agreements.

The report, in my opinion, would have been better if the above issues had been addressed and these or similar recommendations made. It is, nevertheless, a useful document in its present form, and should receive serious consideration.

Laurence I. Moss
Consultant on Energy-Environment; Chairman, Federal Energy Administration's Environment Advisory Committee; Former Executive Secretary, Committee on Public Engineering Policy, National Academy of Sciences; Former President, Sierra Club.

COMMENT, by John Palfrey

While I agree with most of the recommendations and conclusions of this excellent report, I consider that a separate comment is needed on two points.

First, I believe the government should abandon the approach of the Nuclear Fuel Assurance Bill, in which the government assumes the risks that would otherwise discourage private initiatives in building a fourth gaseous diffusion plant, and go ahead and build the plant itself, even at this late date. We are fooling ourselves if we regard uranium enrichment, at least by means of gaseous diffusion, as a conventional private investment opportunity. The initial investment is too great, the period of negative cash flow too long, the prospect of profits too speculative and too remote. More important, the pursuit of "privatization" over the past six years has inflicted incalculable damage on our non-proliferation strategy, because of the delays it imposed on upgrading and enlarging our enrichment capabilities to provide an assured long-term supply of enriched fuel for the domestic and foreign market, while we waited around for industry to complete its studies on private take-over of existing diffusion plants and on private construction of a fourth plant.

We should proceed at once with the government construction of a fourth plant. We should invite the foreign investment participation of advanced countries such as Japan, as well as of Third World countries to enable them to acquire an assured supply of enriched fuel for the future. (This approach would not exclude investment by private industry in the plant, if it so decided.)

Second, at the same time that an accelerated program is launched for the exploration and extraction of additional uranium ores (recommended by the Report), the government should make the decision to defer the licensing of plutonium as a supplementary fuel for current reactors, and concentrate on reprocessing technology primarily as it relates to the supply of plutonium as a primary fuel for breeder reactors, when their commercial feasibility is demonstrated. Such a decision would be justified on the domestic merits of the case, since reprocessing and plutonium recycle are now regarded as marginal commercial ventures at best, and are accompanied by unresolved environmental and safeguards issues.

The decisive consideration, however, would be the paramount objective of "assuring the common defense and security", by reinforcing our efforts to prevent the spread of nuclear weapons. Such a decision by the US, as the most advanced nuclear power country in the world, would greatly strengthen its position among the principal suppliers and Third World countries in opposing the sale of national reprocessing plants to nations with small nuclear programs, where reprocessing would be still more premature, and in persuading the other supplier states to adopt the same position.

If successful, this initiative might provide an interval of several decades in which to establish a genuinely effective system of international controls over the back end of the nuclear fuel cycle.

John Gorham Palfrey
Professor of Law, Columbia University; Former Commissioner, Atomic Energy Commission.

THE ATLANTIC COUNCIL OF THE UNITED STATES

The Atlantic Council, established fifteen years ago, seeks to promote closer mutually advantageous ties between Western Europe, North America, Japan, Australia and New Zealand. The objective is greater security and more effective harmonization of economic, monetary, energy and resource policies for the benefit of the individual in his personal, business, financial and other relations across national boundaries. These varied and complex relationships have been and will continue to be central to the major economic and political developments which affect our international integrity and domestic well-being.

In an increasingly interdependent world where "foreign" policy is ever more closely intertwined with "domestic" policies, there is a clear need for both official and private consideration of means of dealing with problems which transcend national frontiers. The Atlantic Council is a unique non-governmental, bi-partisan, tax-exempt, educational, citizens' organization. It conducts its programs to promote understanding of major international security, political and economic problems, foster informed public debate on these issues, and make substantive policy recommendations to both the Executive and Legislative branches of the US Government, as well as to the appropriate key international organizations.

The Board of Directors of the Atlantic Council is composed of some one hundred prominent leaders and experts in business, finance, labor and education, together with former senior government officials. Their names are listed on the following page.

The Atlantic Council of the United States

BOARD OF DIRECTORS